AF247474

OZ NANO
03

Proceedings of the Asia Pacific Nanotechnology Forum 2003

OZ NANO 03

Cairns, Australia   19–21 November 2003

Jurgen Schulte

Asia Pacific Nanotechnology Forum

World Scientific

NEW JERSEY · LONDON · SINGAPORE · BEIJING · SHANGHAI · HONG KONG · TAIPEI · CHENNAI

Published by

World Scientific Publishing Co. Pte. Ltd.
5 Toh Tuck Link, Singapore 596224
USA office: Suite 202, 1060 Main Street, River Edge, NJ 07661
UK office: 57 Shelton Street, Covent Garden, London WC2H 9HE

British Library Cataloguing-in-Publication Data
A catalogue record for this book is available from the British Library.

Oz NANO 03
Proceedings of the Asia Pacific Nanotechnology Forum 2003

ISBN 981-238-862-1

Printed in Singapore by World Scientific Printers (S) Pte Ltd

CONTENTS

Nanotechnology

Policy

Investment

Human Resources Development

INTRODUCTION

NANOTECHNOLOGY TRENDS IN ASIA PACIFIC

JURGEN SCHULTE

Asia Pacific Nanotechnology Forum
Sydney, NSW 1430, Australia

In 2001 President Clinton brought world wide attention to Nanotechnology with the budget approval for the first US National Nanotechnology Initiative. The initial budget allocated for Nanotechnology in 2001 was (USD) $422 million, which set a clear message about the relevance of Nanotechnology to economic growth as well as its strategic importance to national security. Three years later, in December 2003, President Bush signed the 21st Century Nanotechnology Research and Development Act, which allocated a budget of $849 million for the NNI, almost doubling the initial budget from 2001. After the clear message of commitment of the United States to Nanotechnology in 2001 governments around the world, and in particular in Asia, reassessed their current national Nanotechnology policies and started to develop their own long term position in Nanotechnology. Since then government investments into Nanotechnology has increase to over $3 billion world wide in 2003 with almost half of that investment placed in Asia.

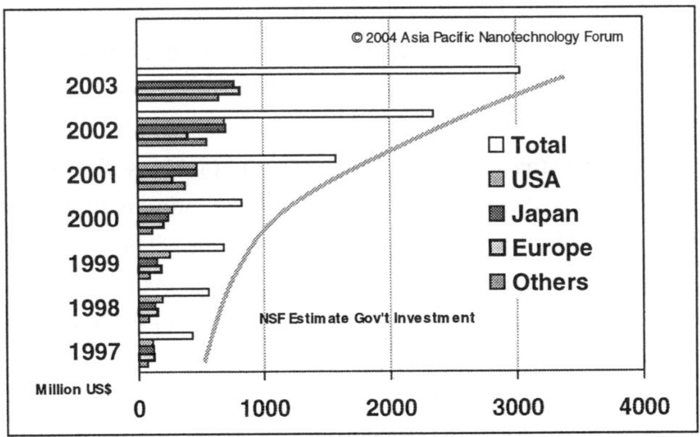

Figure 1. World wide government investment into Nanotechnology.

A number of State Governments have started to implement their Nanotechnology support in addition to the already existing, substantial national government funding, bringing the total public funding for Nanotechnology up to an estimated total of $4 billion. Investment by industry alone was estimated above $1 billion in 2003 (APNF 2003).

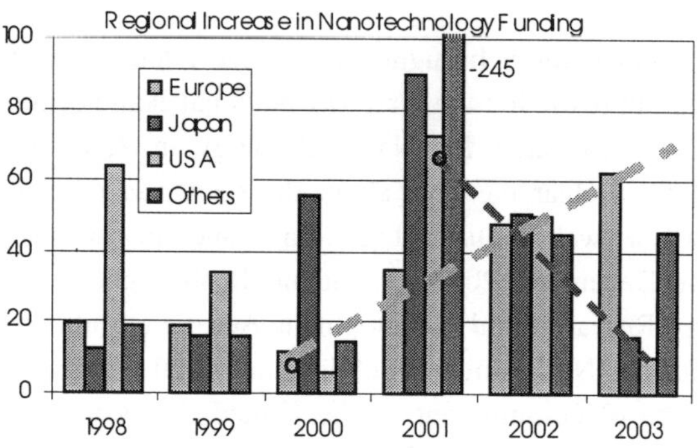

Figure 2. Yearly increase of government Nanotechnology funding by region. Note, the funding for Nanotechnology in Western Europe increased by 245% in 2001.

Figure 1 shows the public, national funding over the past over the past seven years as estimated by the US National Science Foundation. Since 2001 an average of about 20-30% of national funding has been matched by state (province) funding, and industry funding is estimated to exceed government funding except for the smaller Asian development countries. So, overall it is save to say that the total investment into Nanotechnology today is likely to be twice as much as the public funding by governments alone. The yearly increase of government funding for Nanotechnology has slowed over the past three years reflecting both cautious national budgeting and the additional financial input coming from private investment as the first R&D spin offs and products are entering the market space. In Europe, additional public investment in Nanotechnology is increasing at staggering rate (Figure 2) every year since the year 2000. The rest of Asia is used to long-term national plans raging from 5 to 10 years, which requires considerable planning ahead including major infrastructure building such as roads and buildings. China's investment commitment for the five years plan ending 2005 is about $280 million, for Korea's ten years plan ending 2010 it is $2 billion, and $600 million for Taiwan's five year plan ending 2007. Malaysia allocated 9% ($23 million) of its 8th five years Plan Intensification of Priority Research Areas ending 2008 for Nanotechnology and Precision Engineering. Thailand is earmarking $25 million for Nanotechnology for the five years period ending 2008. Australia has identified Nanotechnology as one of its four funding areas of national priority. Asia's public investment in Nanotechnology is now surpassing that of all western competitors together. The funding figures for Nanotechnology show how important it has become to take a stake in Nanotechnology early on and how important it has become to secure fundamental, core intellectual property to secure a future market share in Nanotechnology. It is important to observe that the dollar value of Nanotechnology investment is Asia has a much better turnaround per dollar than its counterpart in USA and Europe due to relatively low cost of highly skilled labor and due to ready access to a very large manufacturing industry.

At this year's Annual Conference of the Asia Pacific Nanotechnology Forum leading policy makers from Asia, Europe, and USA presented reports on outcomes of their respective initiatives, their future directions, and opportunities for others to participate. Venture Capitalists and large corporations highlighted what they are expecting from an emerging technology as diverse as Nanotechnology, and experts in Nanotechnology education discussed educational development issues addressing the human resources requirements in Nanotechnology.

All scientific papers in The Proceedings of the 2nd Annual Conference of the Asia Pacific Nanotechnology Forum are peer reviewed; contributions from the investment community and government policy makers have been printed as presented.

The Asia Pacific Nanotechnology Forum evolved over a number of years from discussions with Nanotechnology Entrepreneurs and Thought Leaders in Nanotechnology in Industry, Government, and Academia. The International Nanotechnology Showcasing Exhibition at the 2nd Annual Conference of the Asia Pacific Nanotechnology Forum, APNF 2003, was the first Nanotechnology exhibition of its kind in Australia.

The Nanotechnology Showcasing Exhibition at APNF 2003 "Oz Nano 03" is a result of the culmination of strategic initiatives which gained their critical momentum in the Australian Nanotechnology community at the joint DISR-CSIRO workshop in March 2001, where for the first time industry, government and representatives from the research community came together to create a "blueprint" for Australian Nanotechnology. On 24 June 2002, the *Australian National Nanotechnology Network* was formed to link the breadth of activity around the country in science (universities, the Australian Research Council, Commonwealth Scientific Industrial Research Organization), government (Federal & State), industry (multi-sector), investment groups, international links, and social impact groups, to ensure the fastest transfer from science to product and to facilitate a leadership position for Australia.

The Australian Government has recognised Nanotechnology as one of its National Research Priorities and it is not surprising that with this tremendous support from the community and government a number of outstanding Nanotechnology innovations have made it to the commercialisation phase and are now part of a solid platform of Australian emerging Nanotechnology companies. At this year's APNF Conference and International Nanotechnology Showcasing Exhibition at The Hilton Cairns leading Australian Nanotechnology companies have gathered to present their products and services in the Australian Nanotechnology Pavilion.

The commitment and encouragement of the following organizations, and their support in particular to nominate and send their delegates to participate in this year's conference and international Nanotechnology Showcasing Exhibition have made the Forum to one the most important Nanotechnology events in Australia in 2003:

Australian National Nanotechnology Network, Australia
Commonwealth Science & Industrial Research Organisation, Australia
National Research Council Canada, Canada
Chinese Academy of Sciences, China
Engineering Research Center of Nano Science and Technology, China
National Natural Science Foundation of China, China
European Commission, European Union
German Center of Competence in Nano-Scale Analysis, Germany
National Institute of Advanced Industrial Science and Technology, Japan
National Institute for Materials Science, Japan
MacDiarmard Institute for Advanced Materials & Nanotechnology, New Zealand
Industrial Technology Research Institute, Taiwan
National Science Council, Taiwan
National Science and Technology Development Agency, Thailand

APEC Center for Technology Foresight, Thailand
The Institute of Nanotechnology, United Kingdom
National Nanotechnology Initiative, USA
National Science Foundation, USA

The Asia Pacific Nanotechnology Forum is grateful to the APNF 2003 "Oz Nano 03" lead sponsors and their generous support throughout the preparation of this event:

Thought Expert Sponsors
InvestAustralia
Thomson Scientific (UK/USA)
The University of Queensland

Thought Catalyst Sponsors
The ARC Special Research Centre for Applied Philosophy and Public Ethics (CAPPE)
The Australian Department of Education Science & Technology

Thought Network Sponsor
The Australian Technology Park (Innovations)

The International Nanotechnology Showcasing Exhibition at APNF 2003 "Oz Nano 03" received generous support through the Australian Innovation Access Program and the
Forefront Thought Leader Sponsors
The Australian Department of Education Science & Technology
The Australian Department of Industry, Tourism, and Resources.

Dr Jurgen Schulte
Executive Director
Asia Pacific Nanotechnology Forum

NANOTECHNOLOGY
FUNDING AND R&D ACTIVITIES IN JAPAN

KAZUNOBU TANAKA

National Institute of Advanced Industrial science and technology (AIST)
(Presented at APNF 2003 " 19-21 November, 2003, Cairns, Australia)

The first part describes the R&D Policy of the Council of Science and Technology Policy, Japan (CSTP) with an emphasis on Nanotechnology, its research funding, several nanotechnology plans and projects running in METI, MEXT and other Ministries. In the second part, recent R&D activities in National Institute of Advanced Science and Technology (AIST), METI, are presented, which will include CNT nanotechnology, spintronics, nano inkjet technology, and MIRAI Project for LSI Road Map.

1. Japan's S&T Basic Plan

New 5-year S&T Basic Plan started in 2001, which was set up by the Council for Science and Technology Policy (CSTP), Cabinet Office of Japanese Government. This plan (FY 2001-2005) includes: (1) Total \24T (~US$220B) government R&D investment in 5 years, (2) strategic prioritization on four big fields; Life Science, Information Technology, Environment, and Nanotechnology / Materials . In particular, funding to Nanotechnology has been strongly emphasized for the past several years, increasing from US$300M (FY2000) to US$510M (FY2001), and US$740M (FY2002).

Six government departments carry out nanotechnology-related national projects in Japan. These are the Ministry of Education, Science, Sports and Culture (MEXT), the Ministry of Economy, Trade and Industry (METI), the Ministry of Public Management, Home Affairs, Posts and Telecommunications, the Ministry of Health, Labor and Welfare, the Ministry of Agriculture, Forestry and Fisheries, and the Ministry of the Environment. The total nanotechnology-related budget for FY2003 is expected to increase more than the previous year's amount of \81.6B (US$740M).

Up to now, most was allocated to MEXT and METI, for example 97.5% of the FY2001 nanotechnology budget for both ministries. The Nanotechnology / materials R&D promotion Project Team (NTPT) of the Council for Science and Technology Policy (CSTP) is trying to strengthen links among government sectors and agencies by organizing inter-ministry collaborative projects that are expected to result in more efficient national R&D investments and research integration. Drug Delivery System (DDS) and Bio-medical nanotechnology projects will start in FY2004 as a main part of inter-ministry collaborative projects.

Under New Research Promotion Program including so-called CREST and ERATO, and Nanotechnology Support Project, MEXT is currently funding nanoscience basic research themes, providing nanotechnology facilities open for users, and constructing nanotechnology network. Total funding of MEXT amounts to around \50.5 billion (US$450M) for FY2003.

METI, on the other hand, focuses on nanotechnology aiming at industrialization. For example, METI assisted to organize the Nanotechnology Business Creation Initiative (NBCI) quite recently in 2003, under which strong interaction among industry, government and academia is encouraged. METI's projects such as those in NEDO Nanotechnology Program also explore a variety of new application fields of nanotechnology. Total funding planned is around \62.4 billion (US$560M) for FY 2003.

2.   AIST Activities Leading Nanotechnology

MITI and its national institutes were reorganized into New METI and new AIST in 2001. As a result, 15 national institutes were integrated into a single independent administrative institution, AIST, while maintaining a close linkage with METI's industrial policy. Now new AIST has become the largest national institute in Japan, staffed with about 7000 people in total, including 2500 permanent researchers. Annual research budget amounts to around US$ 690M including personnel expenditures.

AIST covers a wide range of research fields such as life science, electronics, IT, materials, chemistry, environment, geology, machinery and metrology. Therefore, current nanotechnology programs in AIST include (1) Nano-material, processing, (2) Nano-manufacturing, (3) Advanced semiconductor device project, (4) New spintronics, materials and devices, (5) computational nanoscience, etc.

As for Nanomaterial and processing, carbon nanotubes (CNT) have been investigated with respect to a variety of applications; (a) multi-walled CNT for a field emission emitter under the collaboration with Mie University (Advanced Carbon Materials R.C.), (b) CNT quantum effect transistor, i.e., a room-temperature single electron transistor using a CNT channel prepared through some defect-induced process (Nanotechnology Research Institute), and so on.

A novel nano-manufacturing process has been developed using a new type of ink-jet technology, where ultra fine wiring with a line width less than 3 microns was achieved by direct writing in air atmospheric and desktop factory. Dot size is 1/1000 of current ink-jet printer technology (Nanotechnology Research Institute). Innovative MEMS business-support program (METI) is also going on in AIST (Mechanical Systems Engineering Institute).

The biggest advanced semiconductor device project started in FY 2001, called Millennium Research for Advanced Information Technology (MIRAI) Project, focusing on state-of-the-art semiconductor technology for realizing a 50-70nm node and beyond for SoC by the cooperation of industry, academia and government. More than 100 researchers and annual budget exceeding US$30M are invested (Advanced Semiconductor R.C.).

As for new spintronics and materials, we have two streams of current activities: (a) MRAM with a single-crystal-junction TMR (tunneling magnetic resistance) has been developed using a high-quality Al2O3 barrier. Resonant spin transistor using this barrier was also prepared and oscillation of MR was clearly observed (Nanoelectronics Research Institute). New type of transition metal oxides have been investigated for future spintronics based on novel concept of solid-state physics. These materials, belonging to the family of a strongly correlated electron system, and their basic device technology have been intensively studied towards spintronics devices and new ultrafast optical switching devices (Correlated Electron R.C.).

AIST has been a world-leading pioneer in nanotechnology through pursuing several nanotechnology-related projects such as the Atom Technology Project

(US\$ 220M for 10 years, FY 1992-FY 2001), accumulating a range of results and know-how. In its history, AIST has made it clear that nanotechnology is not only for the hi-tech world but also extends to energy-saving and environment friendly technologies, new biotechnologies, and innovative production processes that result in high-quality component materials. Based on the research results achieved to date, AIST will continue to aim for further breakthroughs contributing to nanotechnology industry.

THE EUROPEAN UNION'S 6TH FRAMEWORK PROGRAMME: RESEARCH ON NANOSCIENCES AND NANOTECHNOLOGIES*

NICHOLAS HARTLEY

*Unit G-1 " Policy aspects", Industrial Research
Research Directorate-General, European Commission, Brussels*

Keywords: nanotechnology, nanosciences, nanotechnologies, European research, Sixth Framework Programme, international co-operation, industrial research, education, social acceptance, sustainable development

1. Introduction

The European Community's multi-annual Framework Programmes for Research and Technological Development provide funding for research carried out at European level.

For the past several years the European Commission (on behalf of the European Community) has supported a significant portfolio of nanosciences and nanotechnologies-related projects. Already in the 4th Framework Programme (1994 - 1998), some 80 projects involving nanotechnology were funded. In the 5th Framework Programme, (1998 - 2002) the estimated funding level rose to about 45 M€/year. The overall project portfolio is very wide in scope, encompassing for example nano-electronic devices, giant magneto-resistance, carbon nano-tubes, bio-sensors, molecular diagnostics, nano-composite materials and atomic force microscopes.

In the case of the current 6th Framework Programme (2003 - 2006), nanosciences and nanotechnologies are a priority. The role played by the Union's initiatives is of paramount importance with respect to the total public funding for nanotechnology activities in Europe. In broad terms, we can estimate the grand total for the European investment in nano-research being now of the order of almost 700M € per year.

The 6th Framework Programme has been tailored to help structure European research in a more efficient and less fragmented way and to cope with the strategic objectives set out by the EU Heads of State and Governments in Lisbon in 2000. This should enable Europe to become the most dynamic and competitive knowledge-based economy of the World within the next 10 years. The commitment is clear: towards sustainable development.

The twofold transition towards knowledge-based society and sustainable development demands new paradigms of production and consumption. There is a need to move from resource-based approaches towards more knowledge-based ones, from quantity to quality, and from mass produced single-use products to new concepts of higher added value, eco-efficient and sustainable products, processes and services.

Materials sciences and nanotechnology allow us now to add value and "intelligence" to materials, components and systems. This will offer great benefits to society and boost the shift from production-oriented approach to a use-and-performance-driven industrial society. Novel activities, and the new generation of high-tech industries associated with these, are showing up on the market. The shift from labour-intensive to brain-intensive operations modifies jobs and the skills required of engineers and the workforce.

To take the benefits from the opportunities afforded by nanotechnology, challenges must be met. There are challenges at a scientific and technical level: to increase our basic understanding of the nano-world; to create new materials, devices and processes; to establish new tools and techniques for industrial manufacture. There are also many challenges at a structural level: to maximise the efficiency of publicly funded pre-competitive research; to educate the next generation of students; to enhance the societal awareness and perception; to use standardisation for increased efficiency; and so on. These challenges call for the formation of Europe-wide alliances of a significant scale between industry, academia, the financial world and other stakeholders.

Top-quality research in very advanced and promising priority technological areas is vital to catalyse the transformation of European industry and society. Nanosciences and nanotechnologies are one of these priority areas.

2. The European Union's 6th Framework Programme for Research and Technological Development

The 6th Framework Programme is formulated within the context of the European Commission's initiative of the "European Research Area". This initiative addresses the perceived weaknesses of European research and seeks to improve co-ordination between the research programmes of the Member States (including the New Applicants), the Associated States and those implemented by the European Commission. It addresses key elements of European research, such as the scope and scale of projects, support for research to the benefit of small and medium-sized enterprises, the critical questions of scientific education, careers and mobility, and the relationship between scientific endeavour and society.

Annual or biannual work programmes are defined for each Thematic Priority Area. To realise the research, the following instruments are used: Integrated Projects (IP), Networks of Excellence (NE), specific targeted research projects (STREP), co-ordination actions (CA), and specific support actions (SSA).

The majority of the funding in the 6th Framework Programme is in the form of grants to research projects. Research will be long term and highly challenging, but oriented towards industrial breakthrough applications. Special emphasis is given to education and training to create the required pool of multi-disciplinary skilled personnel, without whom any substantial progress into the knowledge and exploitation at the "nano level" will be impossible, hence also the need to increase awareness for these new scientific and technological challenges.

In addition, it is expected that breakthrough research activities will help encourage and develop dialogue with society and an increased enthusiasm for science. Whenever appropriate, ethical, societal, health, environmental and regulatory issues, in particular metrology aspects, should be addressed.

With the 6th Framework Programme, Europe does not plan to "go it alone". We are ready to gather a critical mass whenever and wherever appropriate in order to achieve our goals more effectively and more rapidly. Indeed, another peculiarity of the 6th Framework Programme is its openness to international co-operation: researchers from virtually all countries of the world can participate. In particular, co-operation is underlined with those countries, which have a bilateral scientific/technical agreement incorporating nanotechnology, such as Australia, the USA, China, and Russia.

In the 6th Framework Programme (www.cordis.lu/fp6) the main funding channels for research projects and networks are a series of eight Thematic Priority Areas, collectively constituting "Focusing and Integrating Community Research". Of further particular interest are also the chapters "Structuring the European Research Area" (concerned specifically with human resources and mobility, research infrastructures and "Science and Society") as well as "Strengthening the Foundations of the European Research Area".

In order to establish and maintain a coherent approach to nanosciences and nanotechnologies, all of the relevant Thematic Priorities and Specific Programmes operate in co-ordination. In particular, a cross-programme section "nanotechnology" on the CORDIS information service (www.cordis.lu/nanotechnology) has been created. Appropriate criteria are established in order to address concerns about overlaps and voids, so to best serve the scientific community and industry.

Focusing and Integrating Community Research
Given the wide-ranging nature of nanotechnologies, it is likely that several thematic priority areas will provide funding for their application, including dedicated support for medium- and small-sized enterprises. Of particular relevance for the development of nanosciences and nanotechnologies is however the Thematic Priority Area 3 "Nanotechnologies and Nanosciences, Knowledge-based Multifunctional Materials, and New Production Processes and Devices"..

2.1. The Thematic Priority Area 3

The Thematic Priority Area 3 "Nanotechnologies and Nanosciences, Knowledge-based Multifunctional Materials and New Production Processes and Devices" includes a wide-ranging programme on nanosciences and nanotechnologies (including nanobiotechnology) research and development activities. The Community public funding for nanotechnology under this programme may be estimated to mobilise total financial resources in the range of one billion € over the four-year period of the 6th Framework Programme. Two global objectives are defined: first, to stimulate the introduction of innovative nanotechnologies in existing industrial sectors; and second, to stimulate breakthroughs, which can lead to entirely new materials, new devices, new products and new industries. The overall approach is an integrative one, where long-term and medium-term aspects within given scientific/technological developments may be pursued as elements within a project. The programme covers essentially all nano-material types and related device and process developments. It includes long term research, research into techniques and instruments, nano-structured structural and functional materials, and related application development.

The primary objective of the Thematic Priority Area 3 is to promote real industrial breakthroughs and innovative applications, based on scientific and technical excellence. This requires a change in emphasis for Community research activities from the short to the longer term and also in innovation, which must move from incremental to breakthrough strategies. In addition, there is increasing evidence that the transformation of industry towards high-added value organisations necessitates real integrated approaches. These can be either "vertical", combining for example materials sciences, information technologies, nanotechnologies, biotechnologies and production technologies, or more "horizontal" combining multi-sectoral interests.

Other Related Thematic Priority Areas
This broad approach means that nanotechnology applications will be investigated under other Thematic Priority Areas across the 6th Framework Programme, such as those dealing with life sciences, information society technologies, aeronautics and space and energy. In particular, within the Thematic Priority Area 2 "Information Society Technologies", nanotechnology relevant activities contribute mainly to two areas, namely "micro, nano and opto-electronics", and "micro and nano technologies, microsystems and displays". In the first case the objective is to reduce cost, increase the performance and improve reconfigurability, scalability and self-adjusting capabilities of electronic components and systems-on-a-chip. In the latter case, the aim is to bring improved intelligence and functionality in "everything",

exploiting the possibility of networking. This may occur in a distributed fashion or localised for improved miniaturisation and portability. Research may also address introducing or upgrading sensing, actuating operations and improving the interaction of every potential item with their surrounding and with individuals, building upon research in micro-, nano, micro-nano-systems and large integration technologies for an increased quality of life.

2.2. Structuring the European Research Area

Different activities of the chapter "Structuring the ERA" within FP 6 are also relevant to nanotechnologies and nanosciences. Within "Human Resources and Mobility" and under the common "Marie Curie" label (MC), host-driven actions such as "MC Research Training Networks" or "MC Host Fellowships", provide support to organisations hosting European and third-country researchers for transnational training and mobility of researchers. All fields are covered including nanotechnologies and nanosciences. In addition, individual-driven actions such as "MC Intra-European Fellowships" will also be supported. Within the Research Infrastructures activity, different schemes of support is implemented such as, for example, "Transnational Access". Where relevant, this activity, which by its nature and means of implementation is applicable to all fields of research and technology, can be implemented in association with the thematic priorities. Moreover, dedicated initiatives related to the societal and ethical implications of nanotechnology are launched within the "Science and Society" initiatives. The website mentioned above (www.cordis.lu/fp6) can provide more detailed information.

2.3. Strengthening the Foundations of the European Research Area

There are efforts to encourage co-ordination activities, using a bottom-up approach, also in the field of nanosciences and nanotechnologies. The objective of the co-ordination at European level is to enhance the complementarity and synergy between Community actions undertaken under the Framework Programme and those of other European research co-operation organisations, for example within the framework of the European Co-operation in the Field of Scientific and Technical Research (COST). COST is a long-standing bottom-up mechanism that facilitates co-ordination and exchange between nationally funded scientists and research teams in a variety of areas related to nanosciences and nanotechnologies, like physics, chemistry, materials, biotechnology, medicine and health. Reinforced co-ordination among the activities of the European Science Foundation, COST and the Framework Programme is sought in areas of common interest.

3. Meeting Challenges through an Integrated approach

An integrated approach should cover consumption patterns so that the complete industrial cycle conforms to society's requirement for sustainability. Particular attention is given to the strong presence and interaction of innovative enterprises, universities and research organisations in research actions. Europe wide networks and projects are required that give research organisations access to new technologies, therefore stimulating implementation of new approaches in most industrial sectors, in particular SME-intensive sectors. A key issue will be to integrate competitiveness, innovation and sustainability into consistent RTD activities. The integration of education and skills development with research activities will play an important role in increasing European knowledge, in particular in nanosciences and new technologies and opening opportunities for industrial applications.

Industry will be one of the major beneficiaries of nanotechnology, approaching it primarily through a "top-down" aspect. Universities and research centres, on the other hand, are also engaged in exploring self-organising and self-assembling routes – rather more "bottom-up". Both ways need huge investment in terms of people, research, infrastructures and financial resources. The more we generate knowledge and understanding at the nanoscale, the quicker we will be able to use these in industrial production for the benefit of all.

In 2010 or 2015, if future forecasts are to be believed, we can expect nanostructured materials and products to generate income of the order of magnitude of one trillion of € (or $): from electronics to telecommunications, from materials science to biotechnology. To achieve this we necessarily need a new generation of some 2 or 3 million engineers, technicians and operators able to innovate and work in the new nanotechnology-based industries. It will be a tough challenge to educate and train all the newly skilled people these figures imply.

It is difficult to foresee the necessary time for implementing this new technological approach. Apart for a new scientific/technical culture, more entrepreneurship, long term capital, responsible behaviour and clear "rules of the game" are required for entering the "nanoworld". Metrology, pre-normative and pre-standardisation issues, and industrial property rights are some examples.

There are many barriers and challenges. A large critical mass is required in terms of both human and material resources. This implies a new way of co-operation being more open, reinforced, transparent and verifiable.

More knowledge means more power. More power demands more responsibility. The potential offered by more control at the nano-level must be matched by an appropriate analysis and control of the possible risks. The education of new players should be underpinned by a sound ethical consciousness and by responsible attitudes. The debate on the ethical and social aspects of nanotechnology is part of a larger dialogue with the public. It may result in regulations that take into account societal values and demands. Launching broad discussions on this matter is, therefore, also a means to guarantee that new products have a market, since they will be built following requests and expectations of the general public.

Sound, science-based information is already overdue. Transparency is essential: What are we doing? How do we intend to use nanotechnology results? Social acceptance and the trust of everybody will depend crucially on these questions. Appropriate "rules of the game" are also needed, on the one hand to secure industrial property rights and the return on the money invested in research and development, and on the other hand to avoid the creation of "technology dens" where dangerous technologies and applications could be developed.

With the research efforts and resources invested, we can expect that nanosciences and nanotechnologies will progress well. But we may see a "nano-divide" with many countries in the world risking to be excluded. There is, therefore, also an ethical priority to grant everybody interested a way of accessing knowledge, since too much inequality here could lead to unacceptable differences in social-economic development among different regions of the planet.

4. Concluding Remarks

The European Commission aims to create a favourable ground for the development of nanosciences and nanotechnologies research. This will be pursued through a focus on pre-competitive but application oriented research, through vertical integration in projects of significant size that comprise both academic and industrial partners, and through a focus on the multi-disciplinary aspects.

The European Union's activities in nanosciences and nanotechnologies are described on the site http://www.euronanoforum2003.org/index_en.php .   A forthcoming event in Trieste, Italy, « EuroNanoForum2003 », scheduled for 9-12 December, will be an opportunity to see at first hand what is going on in the EU and elsewhere.

Public funds devoted to research in Europe come from the Member States and from the European Union (via the Research Framework Programmes). Research funded by the Union, although less in quantity, plays a pilot role and acts as catalyst of a much larger critical mass.

The final aim is not to develop nanotechnology *per se*, but to give impetus to the research in nanosciences and nanotechnologies to serve the needs and demands of a broader public, to underlie industrial competitiveness within sustainable development, and to contribute to other policies of the European Union.

* This paper does not represent any commitment on behalf of the European Commission. Please   refer   to   official   documents   or   see http://europa.eu.int/comm/research/fp6/index_en.html;   http://www.cordis.lu/fp6;   or http://www.cordis.lu/nanotechnology.

NANOTECHNOLOGY VICTORIA

A VEHICLE FOR COMMERCIALIZING NANOTECHNOLOGIES IN VICTORIA

PETER BINKS

Nanotechnology Victoria Ltd, Australia

The major challenge facing any new technology is the mechanism for converting its potential into commercial reality. For nanotechnologies, this challenge is considerable: requiring significant investment in new capital and skills, to support new manufacturing processes and customer propositions. Australia's existing commercial infrastructure is unlikely to be suited to this challenge.

Today the business model for Nanotechnology Victoria, a new organisation responsible for leading the commercialisation of emerging nanotechnologies for Victorian industry, will be presented. Topics covered include:
- The challenges of nanotechnology commercialisation
- Australian technology management and commercialisation issues
- Nanotechnology Victoria's structure and strategy
- Status and conclusion

1. Nanotechnology Commercialisation

It must be recognized that nanotechnology provides a new challenge both scientifically and commercially. It will require a fundamentally different approach to previous technology revolutions. Nanotechnology has been studied for over a decade now, and its features have become apparent:
- It leverages physical properties very different from those in the macro world; disproportionate electrical and optical properties, physical performance, chemical and biological reactivity, together with selectivity to the molecular level

- It requires cross-disciplinary integration: it combines physical and chemical phenomena with biological materials and materials science effects in new ways and for new purposes
- It provides unprecedented commercial challenges, forcing the reinvention of engineering design, manufacturing, metrology, quality control, and integration in order to leverage features at a scale an order of magnitude below that previously managed

In each of these respects, nanotechnology commercialisation is more akin to space science than to the incremental but important technology changes seen over the last hundred years.

These features, together with the opportunities available for step-change improvement in most industries and activities, are driving major effort and investment across the globe in nanotechnology development. Over US$2 billion is being invested in 2003 in the development of the science, and no major economic entity is not participating.

2. Nanotechnology in Australia

Australia, too, is participating in this revolution. However Australia starts with two significant disadvantages in its: (1) technology management mechanisms, and (2) industrial base. Unless recognised and addressed, these may severely limit its opportunity in nanotechnology commercialisation.

Australia's current approach to technology development and management will be inadequate to capture the benefits of new nanotechnologies. This approach is characterized by:

- A fragmented, underscale research base: 39 universities, a 21-Division CSIRO, plus an equivalent effort in the State Departments and the various technology institutes scattered across the country
- Limited long-term infrastructure: a reluctance to invest for a horizon longer than three years
- Lack of commercialization infrastructure: finance, design and manufacturing skills, commercial management skills, policy expertise
- Absence of a common vision: no national nanotechnology initiative, priorities, coordination, or even information sharing

Australia has succeeded in becoming a technologically-sophisticated nation despite these problems, due to the strengths of brilliant and committed individual researchers and research teams.

The second major disadvantage is the weakness of Australia's industrial base. Australia is a world player in those industries where local efforts are more valuable than national coordination, such as:
- Agriculture and food sciences
- Exploration, resource extraction, mineral processing
- Clinical medicine

Australia has limited involvement in those industries where scale is important, where a longer-term investment position is critical, and where a single focused vision is required to get all parties working to the same objectives. Australia has no meaningful players in:
- Microelectronics
- IT and telecommunications
- Chemicals
- Pharmaceutical industries

The advent of the nanotechnology opportunity therefore presents a critical challenge for Australia's technology development community. Exploitation of nanotechnology opportunities requires precisely the attributes neglected in much of Australia's industry development: vision, focus, cooperation, and integration to provide national scale where institutional scale is lacking. Further, almost alone among the developed nations, Australia is attempting to develop nanotechnology platforms for the growth of new enterprises and industries without a microelectronics industry; with no major chemical industry; and with limited manufacturing industry for consumer devices.

3. Nanotechnology Victoria

Victoria has the same issues, arising from the longevity of both its industry and academic institutions. It has led to the design of a new mechanism, which may allow optimization of scarce resources and establishment of a competitive position in nanotechnology to support its industries.

Nanotechnology Victoria was formed by agreement between the State Government of Victoria and a consortium of three of its major Universities (Monash University, Swinburne University of Technology, and RMIT University) and the CSIRO in November 2002. Its formation was driven by the participants' recognition that the current fragmented approach to the development of nanotechnologies was unlikely to provide commercial benefits to Victorian industry, particularly in competition with the major efforts already underway overseas. The Victorian Government therefore committed $12 million of seed capital to the venture, and the founding members pledged to match that funding from their own resources.

Nanotechnology Victoria Ltd has been formed as an incorporated company and operates according to three key principles:
- Strategic focus: alignment with those industries in which Australia can compete
- Collaboration in nanoscience: working to manage fragmented R&D activities as a single integrated research base
- Commercial management: specifically addressing the gaps in the commercialisation chain

In essence, Nanotechnology Victoria is a corporation with three roles: an investment manager aligned with Australian industry sectors; a research broker; and a commercialization coordinator.

The translation of these principles into management practise is as follows:
- *Strategic focus is achieved through the Board's choice of investments.* Significant work has already been undertaken by the Victorian Government, Nanotechnology Victoria, and other entities to identify the areas of highest priority for nanotechnology investment. These are those areas in which Victoria has established or emerging industries, and can draw upon existing skills and infrastructure for the commercialisation of new technologies. There is effectively a "roadmap" for Victorian industry, and for the emergence of nanotechnology-based opportunities. Projects are chosen to have direct alignment with industry opportunity, not technology excitement.
- *Collaboration between the R&D providers is actively managed through funding and the appointment of key personnel.* Nanotechnology Victoria's projects will draw upon the best that the Victorian institutions – the R&D powerhouses – can offer for a particular

technology development. Each of the four founding members as well as other partners will vest in commercial rights to key technologies. Each will offer their research teams, access to their facilities, and to their commercial teams. Key personnel from those institutions will manage these Programs and Projects under direction of Nanotechnology Victoria management. Founding institutions have committed to match the State Government's investment dollar for dollar with no requirement for investment in particular institutions.

- *Independent and skilled commercial management*: Nanotechnology Victoria is set up as an incorporated company, with a CEO drawn from industry, an independent Chair with corporate experience, and a Board able to act independently of its Members. It will focus management attention and investment on addressing the gap between technology development and product commercialisation. It should be emphasized that Nanotechnology Victoria is not a CRC is any guise; it is not a research centre. It has a mandate to be self-sufficient in funding terms in three years time. It will market its capabilities and those of its Members and seek investment and partnerships from a variety of external companies.
- *Nanotechnology Victoria will also contribute to the broader agenda.* It will oversee nanotechnology education and public awareness initiatives in the State, and contribute to appropriate policy development as required by its stakeholders.

4. Implications of the Model

There are significant implications of this approach, particularly around the priorities for industry relationships:

- Nanotechnology Victoria's key initiatives will be in those sectors of scale and growth in Victoria. These include the clinical health sector; the automotive and aerospace components manufacture; agriculture, food manufacture, and the environment; mineral processing; textiles and materials
- There will be sectors which will not be supported at this time, as there is insufficient local industry to allow commercialisation within Australia. These include the silicon-based microelectronics industry, much of the telecommunications and computing industries, and sectors of the pharmaceutical and chemical industries

This is not "picking winners" but is a process of getting alignment right, recognizing the strengths and weaknesses of our current industrial base.

Nanotechnology Victoria will employ a variety of mechanisms for taking new nanotechnologies to market. It will not rely on either licensing of intellectual property or the formation of spin-out companies to generate wealth; rather it will choose mechanisms suited to the industries it supports:

- In sectors such as clinical medicine and pharmaceutical products we will pursue partnerships with established businesses either from Australia or overseas; spin-off companies may be generated but in many cases will be transition vehicles
- Spin-out companies are more attractive in industries such as the agricultural and environmental technology industries and new materials

Much of Nanotechnology Victoria's initial investment may be in enabling infrastructure – both physical and human – rather than research programs. The intent of these investments will be to provide opportunities for the demonstration and prototype production for new technologies.

The research activities managed by Nanotechnology Victoria will also look very different from those managed by traditional research organisations. They will consist of large objective-oriented commercial and research teams which are managed by personnel employed by Nanotechnology Victoria, and not aligned with individual institutions (Universities or the CSIRO).

Finally, Nanotechnology Victoria will act as a focal point for attracting and allocating funds for developments in Victoria. It will actively seek investment in Victorian and Australian nanotechnology from domestic and international funds, and will play a role in directing investment towards development of the nanotechnology infrastructure, as well as commercially-exciting projects.

5. Current status

Nanotechnology Victoria commenced start-up operations in May, and has now established much of the infrastructure required prior to investing in nanotechnology programs and activities. In particular it has:

- Designed a Board comprising Member representatives and independent Directors; this will be announced in December.   A management team is also being assembled.
- Defined the three major Programs in which research, infrastructure, and commercialisation investments will be made. In these areas a number of exciting projects are emerging.
- Relationships with the R&D suppliers – the Universities and the CSIRO - have been resolved through a Members' Agreement. Similarly, relationships with key partners and Victorian industry players are being developed through a comprehensive marketing strategy.
- Educational and public awareness programs are already underway. Nanotechnology Victoria is working with the Department of Innovation, Industry, and Regional Development in Victoria, The Learning Federation, and the Science Teachers Association of Victoria to introduce innovative packages for nanotechnology education in secondary schools

6. Conclusion

In this paper the major weaknesses in Australia's positioning to participate in nanotechnology developments are identified. Australia faces a significant challenge to capture the benefits of these developments.

For nanotechnology platforms to be made available for Australian industry and society, the critical elements must be:

- leverage of the best resources through collaborative technology management
- focus on and alignment with the sectors of Australian industry best able to capture these opportunities
- application of commercial investment and management skills

Nanotechnology Victoria represents a new model for the commercialisation of nanotechnology research in Australia, which embodies all of these elements.

PRE-SEED INVESTMENT IN AUSTRALIAN NANOTECHNOLOGY

GREG SMITH

Co-Founder, SciVentures Investments Pty. Ltd.,
Level 1, 159 Dorcas Street, South Melbourne, Victoria, 3205, AUSTRALIA

Chair, NANO MNRF, c/- Madsen Building, University of Sydney, NSW 2006

SciVentures sees its role as one of transforming Australian research outcomes into globally oriented start-up companies through early stage investment and management support. It sets out to achieve this by initially focusing on the development of the potential competitive IP strategic position of the technology opportunity and characterizing its likely ability to serve a differentiated global market. Successful completion of these steps leads to the creation of an investible business with a strategic business plan, key executives and a business building. Meanwhile, ongoing research, technology development and early prototyping/trials with lead customers/partners continue as part of the pre-seed development process. In short, the role of pre-seed investment is to complete all steps required for the opportunity to become an attractive next round investment.[1] The author has personally been associated with the commercialization of a number of nanotechnology opportunities in major USA-based corporations. Learnings are described from two particular experiences.

1. Australian Nanotechnology

1.1. *Present Status*

While Australian biotechnology developments are recognized internationally with "Australia punching above its weight", the same hasn't been said for Australian nanotechnology, except perhaps for certain advanced materials developments.[2]

With that in mind, a 2002 DITR-sponsored study on nanotechnology diffusion advocated a future materials emphasis for Australian nanotechnology. Thus,

---

[1] See http://www.sciventures.com.au/ for details.

[2] "Scoping of Nanotechnology Technology Diffusion", DITR, Jan 2002. Prepared for DITR by Ernst & Young, Freehills Technology Services & Howard Partners.

various applications potentials ranging from metals & metal oxides, carbon based structures, to biopolymers, catalysis, conductors/semiconductors, and electromagnets were cited.

However, SciVentures believes that a single focus on materials may not be optimal for nanotechnology R&D investment. The greater part of the value provided by coatings or materials, particularly at a sub-micron scale, is in their use in applications and systems. Generally, little material is required per unit of application for a nanotechnology and the value offered relies more on systems technology and its application, than in a so-called "materials play". With this in mind, we believe that to effectively deploy nanotechnology R&D outcomes, needed systems technologies may require significant partnering with global industry majors.

SciVentures also believes that bio-nanotechnology applications can be built successfully from Australia's biotechnology strength. Australian nanotech. oriented start-up companies in drug delivery, monitoring devices, biochips, diagnostics and sensors are evident.[3] Additionally, it appears that other applications in semi-conductors, lithography and laser applications, although not an internationally well-recognized competitive strength, also may offer key targets for technology start-ups or licensing from Australian nanotechnology.

Thus, SciVentures' strategy is to invest across a broad base of nanotechnologies.

1.2. *Characteristics of Potentially Attractive Start-up Companies*

Clearly, any attractive investment opportunity from nanotechnology needs to offer the potential of developing a strong and competitive intellectual property position in an area where a significant market is likely to emerge.

2. From a Research Outcome to a Pre-Seed Investment

SciVentures (and the other pre-seed funds) have been operating for around a year. The SciVentures Pre-Seed Fund managers have reviewed approximately 200 opportunities, but have only invested in a few to date. The reasons for this are straightforward.

---

[3] ATIP Australia: "Nanotechnology in Australia" April 2003. Purchased report, Asian Technology Information Program (ATIP).

It takes an unusual investor to solely fund ongoing research, however promising that work may be. Normally, an initial investor's goal will be to see that, from successful research outcomes, a business can be built through the development of interactions with prospective customers or partners, and from market research which defines a differentiated, globally-competitive business model for the opportunity. Following that, development of the business and a business plan will proceed to successfully achieve the objectives of the business model. Clearly, it is these steps that can optimize the potential for a positive investment outcome for the first round investor.

The initial investor knows very well that, without these steps being successfully completed, and without key executives available and focused on taking the business beyond this first investment phase, that a next investment round will be either unlikely, or perhaps, will be highly dilutive for the initial investment.

Therefore, before first seeking support from an initial investor, the research team/inventors/founders and their advisors should consider developing some form of preliminary work plan for the sought investment. This should be designed around the likely requirements of the targeted investor. Obviously this will include the completion of R&D, and the filing of patents for any inventions as a high priority. It will also address all recognized steps needed to build the start-up business.

3. Some Past Nanotech Commercialization Examples

During the 1990s, the author managed a number of nanotechnology commercialization activities during his employment with AlliedSignal Inc. and later with Alcoa Inc. He managed corporate research laboratories in the USA for both of these companies.

3.1. *Nanoclay-Nylon 6 compounds*

The polymerization of caprolactam intercalated into montmorillonite clay nanoparticles had been patented by Japanese companies by the early 1990s and shown to offer differentiated nylon 6 engineering plastics performance. At AlliedSignal, an extrusion compounding approach to a similar nanocomposite was considered as an attractive patentable alternative, both economically and for manufacturing flexibility. However, the elimination of any residual aggregates during compound extrusion remained a continuing challenge. While the extrusion concept proved intriguing, and early materials with interesting properties were demonstrated, the achievement of reliable product performance after extrusion compounding took considerable time and invention. Only once this was convincingly demonstrated did the relevant AlliedSignal business group become willing to consider the technology seriously; preferring to consider an external licensing approach up until that point.

3.2. *Diffractive coloring in Fibres and Plastics*

Plastics and fibres are usually pigmented using organic compounds. However, these pigments can change colour gradually under UV exposure or through ozone aging. Additionally, the pigment may be bleached (e.g.) should a stain in a carpet need to be treated with a detergent or bleach. With this in mind, the novel approach of light scattering of sub-micron particles of appropriate size (perhaps on the large size range for nanotechnology) was developed. For particles of certain near mono-disperse particle size, permanent colour in plastics and in fibres was demonstrated for a number of colours, and the concept was patented. Subsequently, endeavours to find a suitably broad range of narrow particle size distribution particles continued with strong support from the corporation's business development leaders who saw this as a very promising, differentiating technology for the company's various plastics and fibres markets. Even shimmering interference colours (so-called "butterfly" colours) were generated, however, the cost effective availability of a broad array of narrow particle size materials across the desired particle size range remained an unresolved challenge.

4. Concluding Remarks

In the above two examples, on the one hand, the circumspection of a business unit client led to a slow but eventually successful technology commercialization as key variables to be overcome for success were kept as a point of focus for the research team. In the second example, a highly attractive market potential led to broad enthusiasm across the research/business development team for several markets. Unfortunately, this enthusiasm masked a killer variable for the technology. While the second technology's potential remains attractive today, the team was clearly addressing a research target rather than a technology solution. The net result was no commercial impact.

In summary, attractive technology opportunities for pre-seed investment will need to be able to demonstrate their potential to develop strong and defendable IP positions. However, even before first investment, they will also have one or more targeted market application(s) in mind, and perhaps some thoughts about how the target(s) might be addressed. The role of pre-seed investment will be to qualify, modify and/or revise these potentials, given the investor's business knowledge and networks, and to help the nanotechnology business team drive through to the point of next round investment readiness.

IN THE FAST LANE OF INNOVATION

BRONTE PRICE

TAFE Industry Partnership Centre, Bay 6, Suite G11, Locomotive Workshop
Australian Technology Park
Garden Street Eveleigh NSW 1430, Australia

While many successful partnerships develop between vocational education and training (VET) providers and emerging industries, challenges remain for the VET sector as a whole to keep up with the fast pace of change. This paper explores some of these challenges.

1. Introduction

The major public training provider in Australia, Technical And Further Education (TAFE) is a significant player in the VET sector. The vast majority of courses delivered by TAFE are work-related. They form part of the Australian national training framework that has evolved over the past decade. This system is competency-based, and courses that are part of that system have been developed in consultation with industry. In 2002, throughout Australia, TAFE delivered courses to more than 1.7 million students, many of whom attended courses at one of 1461 TAFE colleges associated with the 85 TAFE Institutes in Australia.

In the Australian state of New South Wales (NSW), in 2002, more than 570 000 students studied TAFE courses through the 131 TAFE NSW colleges that form the 12 TAFE NSW Institutes. Most TAFE NSW courses have credit transfer arrangements with Australian universities and many of these courses are delivered not only on campus but also in the workplace.

The size and extent of TAFE has led to it having the following strengths:

- tradition of vocational and technical skill development
- substantial community reach at the local level

- extensive national infrastructure
- solid relationships with industry and education providers
- flexibility and diversity that enable it to address complex and multiple needs.

2. Innovation, Emerging Technologies and Emerging Industries:

An important factor in Australia's capacity to create wealth and compete in the global market economy is its ability to capture, develop and implement new and emerging technologies. Emerging technologies such as nanotechnology, biotechnology, bioinformatics, smart card technologies, photonics, information and communication technologies are important in creating new products and services, as well as innovative business models and work practices.

For Australia to keep pace in a global knowledge economy, we need to develop and apply these emerging technologies efficiently and effectively. Emerging technologies and, in particular, emerging industries, have needs that are different from traditional industries that are based on mature technologies. This is of particular relevance to TAFE and the VET sector, which were originally established to meet the training needs of these traditional industries.

Emerging industries are characterised by:

- A rapid pace of growth that demands of their people enormous energies and inputs of time;
- The need to attract and source investment in the form of high risk capital;
- Skills development that crosses traditional skill boundaries and so is outside traditional skill pathways development; and
- A myriad of other requirements associated with building start-up companies.

In addition, there are two types of innovation that most emerging industries acknowledge – incremental and disruptive. Incremental innovation is about the application of an innovative technique to an existing product or process or service – a gradual change that improves what already exists.

Disruptive innovation occurs when the application of an idea completely changes an existing product, process or service, rendering the status quo obsolete, or introduces a new one.

The Australian national training framework needs to accommodate the implications of both types of innovation, since emerging industries are also characterised by:

- development of disruptive technologies, products and services;
- development of disruptive business models and vocational settings that place new demands on employees.

Emerging technologies can also change industrial practice quickly. As an emerging technology becomes a routine part of an industry's componentry, new skill requirements emerge.

These industries' requirements of VET are also different, and the TAFE Industry Partnership Centre at the Australian Technology Park (ATP) in Sydney was set up in the recognition of this. Its goals are to enhance the provision of industry-relevant VET programs to these emerging industries and to diffuse information throughout TAFE NSW about the research and development processes of a number of emerging industries. The Centre does this by initially fostering awareness, then brokering schemes of involvement and participation. The business program of the Centre has largely been a process of discovery and development since it provides a key intersection between TAFE and the innovation sector. One of the key successes that has have ensued from the work of the Centre, for example, is a strong alliance between the Australian Photonics Cooperative Research Centre (CRC) and TAFE NSW - Southern Sydney Institute that led to the development of three courses in Photonics at Certificate II, Diploma and Advanced Diploma levels and, later, NSW Board of Studies approval for the Certificate II in Photonics to be a TAFE-delivered VET (TVET) course as part of the Higher School Certificate for secondary school students.

There are major challenges for the VET sector as a whole in meeting the specific needs of emerging industries and technologies. Areas of research lead into commercial production that is characterised by a rapid pace of growth but it is often small companies which are involved in the generation of innovation and emerging technologies. Generally, the people who manage these companies are both cash-poor and time poor. In addition, the vast majority of these companies are small to medium enterprises. VET is not a high priority for them and they do not have the money, time or the will to undertake studies towards a full qualification at a local TAFE college. This is despite the fact that many admit that this attitude potentially has significant consequences for them.

Moreover, many people feel that it is easier to set up strategic partnerships with large companies than it is to work with clusters of companies involved in emerging technologies. And there are several pivotal training issues that apply to emerging industries.

3.   The Thirst for Skills

It is important to recognise that any competencies that managers in emerging industry areas help to develop will most likely be superseded within one to two years, requiring their further input into a continual updating process. This results in a process which requires their considerable involvement, leading to questions about return on investment for small companies. Furthermore, the attainment of competencies and skills alone is not sufficient to add value to these companies: they need emotional wisdom and business knowledge – topics that can quickly add value to them and which are based on genuine working experiences.

A second key issue is that companies that work in emerging industry areas find it difficult to source sufficient workers with the required skills.

Often these companies are run by entrepreneurs who demand of their workers an ability and a commitment to work in teams; an ability to contribute to a culture in which innovation is encouraged; an enthusiasm to learn new things and to share those things with their co-workers; and a passion about innovation and new technologies. These requirements are not always acknowledged and addressed by VET programs.

Furthermore, the employment market in emerging industries is volatile. Large and very fast increases and decreases in the number of people required can occur at any stage of the company's life cycle.

An integral component of this 'time to market' factor is skill development. The challenge here for the VET sector is to meet this type of demand when the initial costs of course (or topic) development are high and the numbers of potential students are initially low.

4. Emerging Industries and Australian Training Packages

Development of new courses in emerging technology areas is sometimes hampered by lengthy approval and accreditation processes in the development of a new Training Package. The units of competency developed can sometimes be redundant before they have been implemented. Clearly, this could easily happen in emerging technologies such as nanotechnology, where the pace of change is so rapid.

Course content about these technologies needs to be embedded rather than to stand alone as distinct qualifications. This particularly makes sense for content about those technologies that are converging with IT, such as nanotechnology. The mainstreaming of nanotechnology will require different skill sets than those we currently identify and teach in TAFE. It follows that the rollout of nanotechnology courses will also require different training content and different ways of delivering that content. This is a major challenge for the VET sector, as it will require the sector as a whole to address the issue of how to keep up with this embedding and how it will ensure its teachers maintain their currency of content knowledge and teaching skills in the field of nanotechnology.

Training Package content also needs to take into account that the small business management training needs of small to medium enterprises working in emerging technologies such as nanotechnology and associated emerging industries are quite different, in some ways, from those of companies of similar size working in traditional industries. Generally, the manager of small retail businesses are unlikely to request training topics such as 'dealing with alliances' or 'sources and options for SME finance' or 'entrepreneurial leadership' and so on.

In addition, emerging technologies are often 'enabling' technologies – that is, they have the capacity to change the whole basis of competition within a sector but have not yet been embodied within a product or process or service. By definition, they will run across a range of industry areas. For this reason, teaching areas will need to be straddled and boundaries will need to be blurred if these training needs are to be met. This is a new challenge for the VET sector.

5. Conclusion

Clearly, then, there is a need for flexible approaches by TAFE NSW in dealing with emerging industries, as the successes and challenges discussed in this paper demonstrate. New skills and new practices, developed through research and development, need to be successfully fed into emerging industries. The specific needs of these industries must be taken into account if greater synergies are to develop between VET and emerging industries, ensuring that these industries maintain and enhance their current key position in the Australian economy. To implement innovative new courses such as those around nanotechnology will require some flexibility on the part of State and Commonwealth government agencies. Traditional approval processes may not be appropriate and new policies may need to be implemented to enable the VET sector to meet the needs of emerging industries.

www.tafensw.edu.au/ipc

Nanoscience
Nanotechnology
Bio-Nanotechnology

Peer reviewed papers.

ELECTROCHEMICAL CHARACTERISTICS OF ORDERED MESOPOROUS CARBON AS ELECTRODE MATERIAL IN LI-ION BATTERY[*]

W. XING

The Nanomaterials Centre, School of Engineering, University of Queensland, Brisbane 4072, AUSTRALIA

Z. F. YAN

State Key Laboratory for Heavy Oil Processing, Key Laboratory of Catalysis, CNPC, Universtiy of Petroleum, Dongying 257061, CHINA

G. Q. LU**

The Nanomaterials Centre, School of Engineering, University of Queensland, Brisbane 4072, AUSTRALIA

R. RYOO

National Creative Research Initiative Center for Functional Nanomaterials and Department of Chemistry, KAIST, Yusung, Taejon, 305-701, Korea

Ordered mesoporous carbon CMK-5 was comprehensively tested for the first time as an electrode material in Lithium ion battery. Electrochemical properties of CMK-5 were studied by galvanostatic cycling and cyclic voltammetry. Results show that the reversible capacity of CMK-5 was 525 mAh/g at the third charge-discharge cycle and that CMK-5 was compatible for quick charge-discharge cycling because of its special mesoporous structure. Of special interest is that the CMK-5 gave no peak on its positive sweep of the cyclic voltammetry, which was different from other known anode materials.

1. Introduction

The rechargeable Li-ion battery (LIB) has been the dominating electrical energy storage means since Sony first introduced it into the market in 1991. It has been widely used in many portable electronics, telecommunication equipment and hopefully in the electric vehicles in the future. Increasing the energy storage density and making it more compatible with high power operations has been the primary targets of researchers in this area. It is well known that graphite is the commercially used anode material in LIB at present. Its capacity is around 372 mAh/g, which corresponds to the theoretical limit of LiC_6. New anode materials with better properties than conventional graphite are being attempted and tested by researchers.

Ordered mesoporous carbon, CMK-5, is a novel nano-carbon that consists of arrays of nanotubes. Since its emergence, it has attracted much attention in electrochemical and energy-related applications [1]. This kind of carbon is formed by using ordered mesoporous silica as templates, the removal of which leaves a partially ordered graphitic framework. The carbon skeleton results from the carbonization of polymerized furfuryl alcohol, which has been shown to have a very high capacity for lithium insertion [2]. Mesoporous carbon CMK-5 possibly combines the high energy capacity and fast lithium ion transfer (compatible with quick charge-discharge operation) in its structure due to its intrinsic carbon properties and its ordered pore structures, respectively.

2.   Experimental

2.1. *Synthesis of mesoporous carbon CMK-5*

Mesoporous carbon was synthesized according to Ryoo's method. Mesoporous silica SBA-15 was employed as template. A suitable amount of furfuryl alcohol was induced into the pore of SBA-15, followed by the polymerization at 95 °C. Mesoporous carbon was obtained after carbonization at 900 °C and removal of the silica template by HF solution. The detailed procedures of CMK-5 synthesis can be found in the literature [1].

2.2. *Electrode preparation and electrochemical tests*

The experimental cell is a two-electrode cell with a lithium metal sheet as a counter electrode. A working electrode was made by spreading a paste mixture of 80 wt% mesoporous carbon, 10 wt% carbon black and 10wt% PVDF (Aldrich) on a 1×1 cm square copper sheet. The active material applied on the electrode was as small as 1 milligram in order to test the compatibility of its properties with quick charge-discharge operations. Both electrodes were separated by a polypropylene non-woven porous sheet, which was soaked with 1 M $LiPF_6$ solution of ethylene carbonate (EC) and diethyl carbonate (DEC) solvent mixture (1:1 volume ratio). Each cell was cycled at a high constant current density of 200 $\mu A/cm^2$ (equal to about 200 mA/g for CMK-5) with cut-off voltages of 0.005 V in discharge and 2.0 V in charge vs. lithium metal. Here, discharge and charge indicated lithium intercalation into and deintercalation from the carbon, respectively. Linear sweep cyclic voltammetry (CV), mainly with a sweep rate of 1 mVs^{-1} was performed at room temperature. A solartron 1480 multistat was used for the galvanostatic cycling and cyclic voltammetry.

3. Results and discussion

3.1. *Materials characterization*

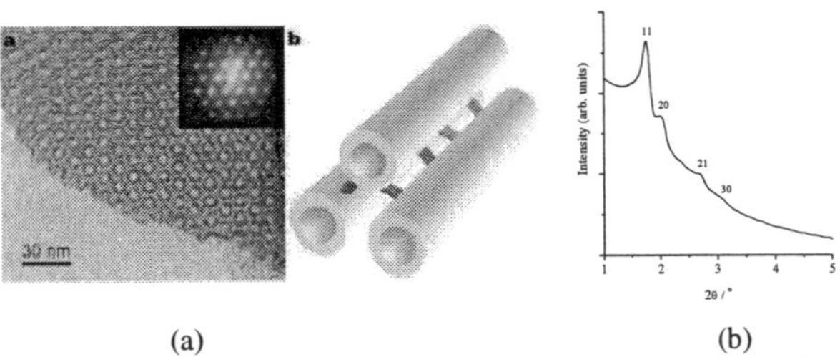

(a)   (b)

Fig. 1 TEM micrograph (a) [1] and small angle XRD spectrum (b) of CMK-5

Mesoporous carbon CMK-5 was constructed with hexagonal arrays of carbon nanotubes or nanopipes as shown by the TEM image and its structure model in Fig. 1 (a) [1]. The CMK-5 was synthesized with SBA-15, of which the pore walls were coated with carbon films. The carbon nanopipes retained the hexagonally ordered arrangement of SBA-15 permanently after carbonization. The wall thickness of the carbon nanopores can be controlled to a certain degree by the amount of source carbon. The inner diameter of carbon tubes was around 3-4 nanometre according to N_2 adsorption analysis. Seen from Fig. 1 (b), the diffraction peaks in its low angle X-ray diffraction spectrum indicates the long-range hexagonal order between mesoporous carbon nanotubes.

3.2. *Electrochemical properties*

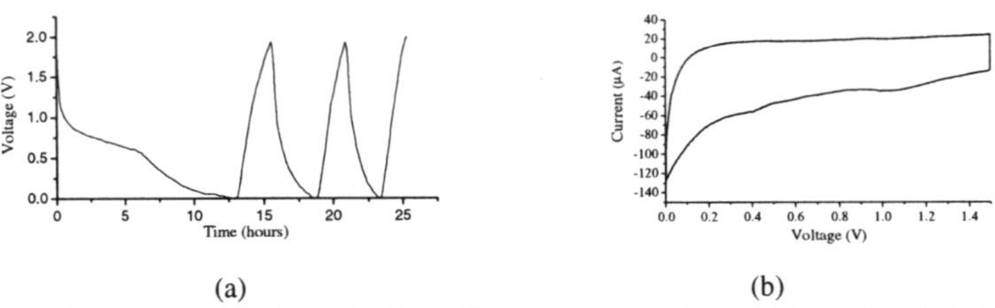

(a)   (b)

Fig. 2 Electrochemical characteristics of mesoporous carbon CMK-5, (a) First three charge-discharge cycles of CMK-5; (b) Cyclic voltammograms of CMK-5 after three cycles. Scan rate 1 mV s^{-1}.

Seen from Fig. 2 (a), the discharge and charge curves of CMK-5 did not present any obvious potential plateau at the low voltage below 0.3 Volts after the first cycle, which was similar to the other non-graphitic carbons. The carbonization temperature for making CMK-5 is low (900 °C) so that the graphitization level of the framework of CMK-5 is very poor. If the carbonization is complete at much higher temperatures such as >1500 °C, CMK-5 would expect to be more graphitic and probably to present a potential plateau at low voltages, which is very significant for its practical application. As seen from Fig. 2(a), the long discharge curve of the first cycle was due to severe irreversible reactions occurring on the large surface of this carbon to form SEI film. During the first three cycles, charge capacities changed from 575 to 600 and 525 mAh/g, respectively. The high charge capacity values show that CMK-5 was compatible for quick charge-discharge operations because of its high surface exposure to electrolyte and its special porous structure favouring lithium transfer during intercalation. As seen from Fig. 2 (b), on the negative sweep of CV, a very small irreversible cathodic peak at about 1.0 volt was presented, which means that the SEI formation process was nearly complete at this time. Of especially interest is that CMK-5 gave no peak on its positive sweep of the CV, which was expected to be due to its special nanostructures. The current density kept almost constant (about 20 μA) on its positive sweep curve. In the case of graphite, the positive curve usually presents a sharp peak at around 0.25 Volts. The level positive sweep curve of CMK-5 indicates that the intercalated lithium ion can move out uniformly during the following de-intercalation process. This indicates that lithium ion possibly redistributed again to the different sites with various energy level through the nanopores after intercalating into CMK-5.

4. Conclusion

Mesoporous carbon CMK-5 was synthesized and applied as electrode material for the first time in the lithium ion battery. XRD study reveals that the synthesized CMK-5 possesses a periodic mesoporous structure. Galvanostatic cycling tests showed that CMK-5 was compatible with quick cycling operations because it can maintain very high capacity under this condition. This property was expected to derive from its unique mesoporous structure. Quick cycling compatibility will make this carbon material very promising in some practical scenarios where quick charge-discharge operations are required.

Acknowledgements

Financial support from the Australian Research Council is gratefully acknowledged.

References

1. S. H. Joo, S. J. Choi, I. Oh, J. Kwak, Z. Liu, O. Terasaki and R. Ryoo, Nature, **412**, 169 (2001).
2. S. J. Wang, Huaxue Yanjiu Yu Yingyong, **13**, 383 (2001).

LIGHT EMISSION FROM SILICON NANOCRYSTALS — SIZE DOES MATTER!

ROBERT G. ELLIMAN

Electronic Materials Engineering Department, Research School of Physical Sciences and Engineering, Australian National University, Canberra, ACT0200, Australia

Silicon nanocrystals exhibit strong room-temperature luminescence as a direct consequence of their small physical dimensions. Moreover, the emission wavelength can be tuned by controlling the nanocrystal size distribution. This paper summarises recent experiments on: a) hydrogen passivation of luminescence-limiting defects, b) the use of optical structures to control the nanocrystal emission spectrum, and c) carrier dynamics in photo-excited nanocrystals.

1. Introduction

Silicon's pre-eminence in high-speed digital electronics does not generally extend to optoelectronics where the demand is for devices that can generate, guide, detect and process optical signals. However, the utility of this ubiquitous material has recently been extended by the observation that porous and nanocrystalline Si exhibit strong room-temperature luminescence[1]. This strong emission is a direct consequence of the nano-scale dimensions of the material and can be tuned to different wavelengths by controlling the nanocrystal size distribution. As a consequence, Si nanocrystals offer an attractive basis for the full integration of electronic and optical functionality in Si-based devices and structures. This paper provides insight into the light emitting properties and applications of Si nanocrystals by briefly describing three current studies undertaken at the ANU, namely: a) improved emission efficiency by hydrogen passivation of non-radiative defects, b) control of emission by the use of optical microcavitites, and c) carrier dynamics and the prospect for optical-gain and lasing in Si-based materials.

2. Experimental

Specific experimental conditions vary for each of the studies reported below but all involve Si nanocrystals produced in SiO_2 by the precipitation of excess Si ($\sim$ 10 at. %). This excess was produced either by plasma-enhanced chemical vapour deposition (PECVD) or from stoichiometric SiO_2 by Si-implantation. Annealing at temperatures around 1100°C for 1 hour was then used to produce nanocrystals with a size distribution in the range 1-5 nm. The photoluminescence emission from these materials was typically in the range 750-900 nm. Specific experimental details can be found in relevant references.

3. Results and Discussion

3.1. *Defect Passivation*

Defects at the Si/SiO_2 interface can act as non-radiative recombination centers thereby reducing the luminescence efficiency of nanocrystals. These can be passivated by hydrogen[2-7] to enhance the luminescence efficiency but optimization of this process requires an understanding of the relevant reaction kinetics. In a detailed study[5,6], time-resolved photoluminescence (PL) measurements were used to determine the defect concentration as a function of annealing time and temperature. These measurements were then modeled using a generalized treatment that included simultaneous hydrogen passivation and hydrogen desorption, as well as a spread in activation energies for each reaction[7]. A summary of the data and a fit of the model are shown in Fig. 1.

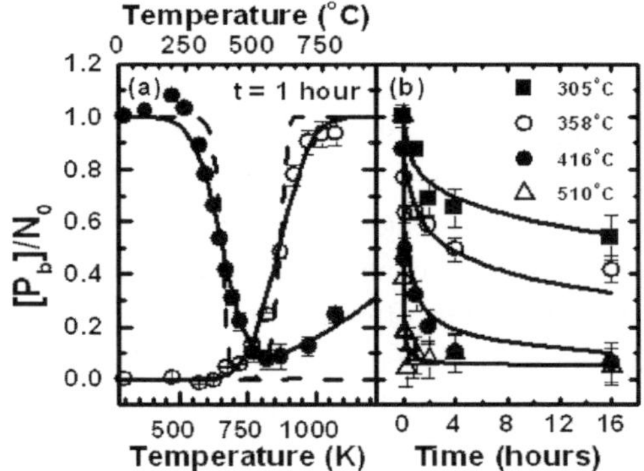

Fig. 1: Normalised defect density from time-resolved PL measurements. a) Isothermal anneals. Solid symbols for annealing in 5% H_2 (passivation), open symbols for annealing in N_2 (desorption). Solid line is model fit, dotted lines are for a simplified model, b) Isochronal anneals in H_2 (passivation).

Values for the reaction-rate parameters were determined from the model[5, 6] and found to be in excellent agreement with values previously determined for paramagnetic Si dangling-bond defects (P_b-type centers) found at planar Si/SiO_2 interfaces[7]; thus supporting the view that non-radiative recombination in Si nanocrystals is dominated by such defects. In addition to improving the understanding of the passivation mechanism this data enabled the determination of an optimal passivation schedule for Si nanocrystals[5, 6].

3.2. *Emission Control*

The spectral emission from an ensemble of Si nanocrystals is inhomogeneously broadened due to the distribution of nanocrystal sizes. For many applications, however, it is desirable to achieve narrow-band emission. This has led to the use of alternative strategies, including the use of optical microcavities in which the Si nanocrystals are embedded between two Bragg-mirror structures (see Fig. 2).

At the ANU, such structures have been fabricated by PECVD of SiO_2, Si_3N_4 and SiO_x layers[8]. The nanocrystals were formed within a Si-rich oxide layer (SiO_x) sandwiched between two parallel distributed Bragg mirrors (DBM) made from alternate SiO_2/Si_3N_4 layers, as shown in Fig. 2. Fig. 3 compares the optical emission from an isolated SiO_2: nanocrystal layer with that from the same layer incorporated in a microcavity structure[8], highlighting the added control afforded by the microcavity structure.

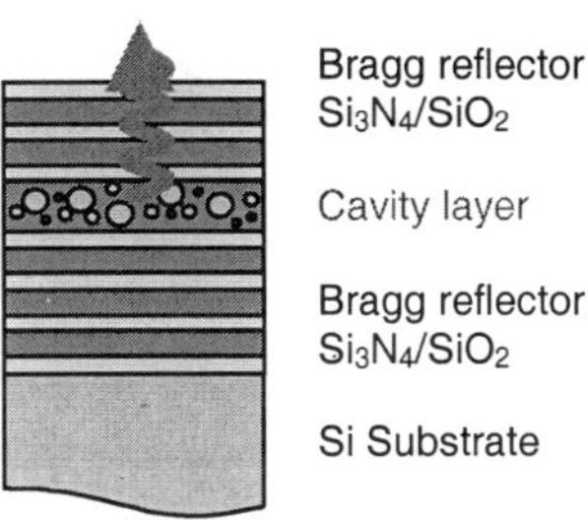

Fig. 2: Schematic diagram of a microcavity structure consisting of an SiO_2:Si nanocrystal layer embedded between two distributed Bragg mirrors.

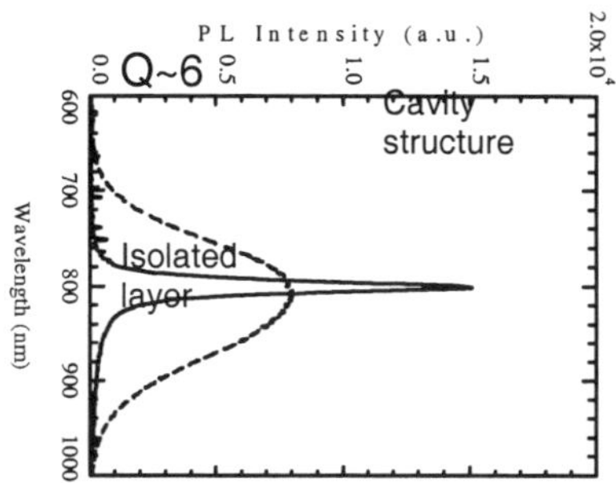

Fig. 3: PL spectra from an isolated SiO_2:Si nanocrystal layer and from the same layer embedded in a microcavity structure (see Fig. 2).

3.3. *Carrier Dynamics*

An exciting and controversial report of optical gain produced by photo-excited silicon nanocrystals was recently published in Nature[9]. In an attempt to verify this claim, optical pump-probe experiments were undertaken[10, 11] in which an optical probe beam (wavelength of 800 nm) was propagated through a slab waveguide containing silicon nanocrystals and monitored as the nanocrystals were optically excited by a high energy density (3.5-3500 $\mu J/cm^2$) pump beam (wavelength 355 nm). This arrangement, shown in Fig. 4, provided many advantages over those previously employed, including improved sensitivity.

Time resolved measurements of the guided probe signal, shown in Fig. 5, show that the probe intensity decreased when the waveguide was subjected to optical pumping –no gain was observed. Indeed, the results were shown to be consistent with an excited state absorption process, in which photo-generated carriers induce 'free' carrier absorption of the probe beam. The model was supported by the fact that the induced absorption lifetime ($\sim$50μs) was the same as that of nanocrystal luminescence –i.e. the lifetime of the carriers. Such experiments[10-12] have raised doubts about the model originally proposed for the gain but the question of whether or not gain can be achieved in the SiO_2: nanocrystal system remains open and is an active area of current research.

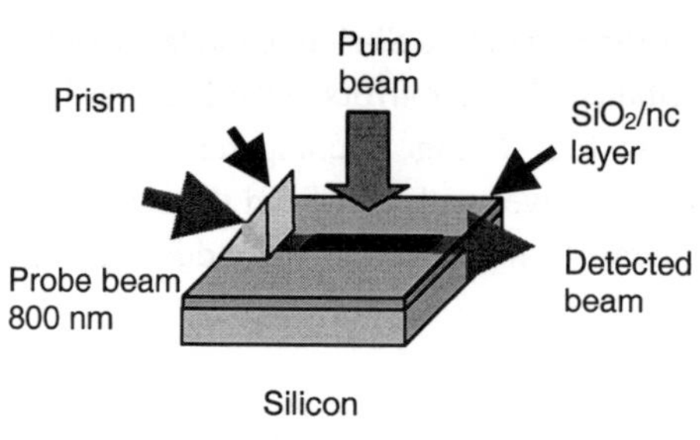

Fig. 4: Schematic diagram of the arrangement used for pump-probe measurements in a waveguide geometry.

Fig. 5: Time-resolved measurement of the guided probe intensity following irradiation with a 25 ns pump pulse.

4. Conclusions

The ability of Si nanostructures to emit light makes them strong candidates for Si-based photonics devices and may enable the full integration of electronic and photonic functionality for specific applications. Selected examples from the ANU were employed to illustrate the interesting physical properties and applications of Si nanocrystals and related structures.

Acknowledgments

Major contributors to the above work include: A. Wilkinson, N. Smith, M. Spooner, T. Walsh, S. Orbons, M. Lederer, M. Samoc and B. Luther-Davies.

References

1. D. Kovalev, et. al., Phys. Stat. Sol., **B215**, 871 (1999)
2. S. Cheylan and R.G. Elliman, Appl. Phys. Lett., **78**, 1225 (2001)
3. S. Cheylan and R.G. Elliman, Nucl. Instr. Meth., **B175-177**, 422 (2001)
4. S. Cheylan and R.G. Elliman, Appl. Phys. Lett., **78**, 1912 (2001)
5. A.R. Wilkinson and R.G. Elliman, Mat. Res.Soc.Symp.Proc., **770**, 81 (2003)
6. A.R. Wilkinson and R.G. Elliman, Phys. Rev., **B68**, 155302 (2003)
7. A. Stesmans, J. Appl. Phys., **88**, 489 (2000)
8. M.G. Spooner, T.M. Walsh, and R.G. Elliman, Mat.Res.Soc.Symp.Proc., **770**, 51 (2003)
9. L. Pavesi, L.D. Negro, et. al., Nature, **408**, 440 (2000)
10. R.G. Elliman, M.J. Lederer, N. Smith, and B. Luther-Davies, Nucl. Instr. Meth., **206**, 427 (2003)
11. N. Smith, M.J. Lederer, M. Samoc, B. Luther-Davies, and R.G. Elliman, Mat.Res.Soc.Symp.Proc., **770**, 63 (2003)
12. A. Polman and R.G. Elliman. *Optical Gain from Silicon -A Critical Perspective*. in *NATO Advanced Research Workshop -Towards a Silicon Laser*. 2002. Trento, Italy: NATO Science Series (Kluwer Dordrecht) (2003)

APPLICATIONS OF PLASMONICS IN MINIATURISED SENSORS

P.R. STODDART, J.B. PEARCE, T.M. BOYCE, A. MAZZOLINI

School of Biophysical Sciences and Electrical Engineering, Swinburne University of Technology, PO Box 218, Hawthorn, Victoria 3122, Australia

Surface plasmon resonances that occur in nanoscale noble metal films and particles are extremely sensitive to the local dielectric environment. As a result, both the surface plasmon resonance (SPR) and the localized surface plasmon resonance (LSPR) have attracted a great deal of interest for diverse biological and chemical sensing applications. One avenue to exploiting the remarkable optical properties of nanoscale metals is to make use of their intrinsic capacity for miniaturization, in combination with the powerful manufacturing capabilities that have been developed in the microelectronics and optical telecommunications industries. In this paper we firstly explore the potential for an SPR moisture sensor based on a gold-on-silicon structure. This configuration is shown to be highly sensitive to the dynamics of surface droplet formation. Secondly, we describe efforts to use the LSPR to generate surface enhanced Raman scattering on the tip of an optical fiber. This requires the preparation of well-defined and stable nanoscale features.

1. Introduction

Mesoscale aggregates of atoms and molecules are particularly attractive for sensing applications, because the physical properties of smaller devices are generally more susceptible to alteration[1]. For example, plasmon resonances that occur in nanoscale metal films and particles are sensitively dependent on the local refractive index, which generally changes if analyte molecules are immobilized on the metal surface. These resonances occur when light of the correct wavelength excites collective oscillations in the plasma of conduction electrons.

The surface plasmon resonance (SPR) may be excited at the interface between two media with dielectric constants of opposite sign, such as a metal and a dielectric. The excitations propagate along the surfaces of the metal, with most of the electromagnetic field concentrated in the dielectric. On the other hand, the localized surface plasmon resonance (LSPR) that occurs in Au, Ag or Cu nanoparticles is non-propagating, but also leads to an enhanced electromagnetic field in the near surface region of the particles. This contributes to the large enhancements found in surface enhanced Raman scattering (SERS)[2]. Therefore nanoscale films and particles present the possibility of highly sensitive sensors with multiple transduction mechanisms.

In order to reduce costs and improve the reliability of plasmonic sensors, we describe here our efforts to apply silicon and optical fiber materials to the miniaturization of plasmonic sensors. Firstly we describe an SPR analog to the well-established chilled mirror dew point hygrometer. Secondly, the challenges of producing an optical fiber SERS sensor are discussed in the context of some recent results.

2. Silicon-based Moisture Sensor

Although a plethora of humidity sensing mechanisms have been proposed, the chilled mirror hygrometer remains one of the most accurate and reliable methods[3]. In this system, a mirror is cooled until dew formation occurs on the surface. A beam of light reflected off the surface is used to monitor the condensation. Although chilled mirror hygrometers can provide accuracies of ±0.2% RH, current devices are relatively expensive and bulky and require careful maintenance of the mirror surface.

In order to address these issues, we have examined an SPR-based moisture sensor. The device uses a variation on the attenuated total reflection method of Kretschmann[4], but is based on a simple silicon configuration. Light from a 1300 nm laser diode is coupled into a 40 x 7 mm section of a gold-coated silicon wafer, through an end window polished at an angle of 16.5° to the surface normal. This angle was based on the expected critical angle for the plasmon resonance, calculated from the known dielectric constants. Silicon is transparent at this wavelength and the wafer is polished on both sides. Therefore any totally internally reflected light will be transmitted to an InGaAs photodetector. However, if the light excites a plasmon resonance in the 45 nm gold layer, the transmitted intensity will be diminished. The SPR system was mounted on a rotation stage to assist in finding the resonance angle. This was found close to the predicted value of 16.5°. From resonance, the transmitted intensity was shown to rise by *ca.* 30% with the presence of water on the gold surface. This was a satisfactory result, given that the polarization direction of the laser diode source was not controlled.

Initial tests were performed by breathing on the gold surface while it was held at room temperature, leading to the results shown in Fig. 1. Here the SPR transmission is compared to the intensity of a second laser beam reflected off the gold surface. Clearly both measurement methods respond very rapidly to the onset of condensation. However, while the reflected intensity recovers slowly as the condensation evaporates, the transmitted intensity displays a more complicated behavior, particularly for longer breaths. Additional transmission measurements, performed while heating the apparatus with a hair dryer, have demonstrated that these effects are not caused by thermal changes or vibrations associated with the breath delivery.

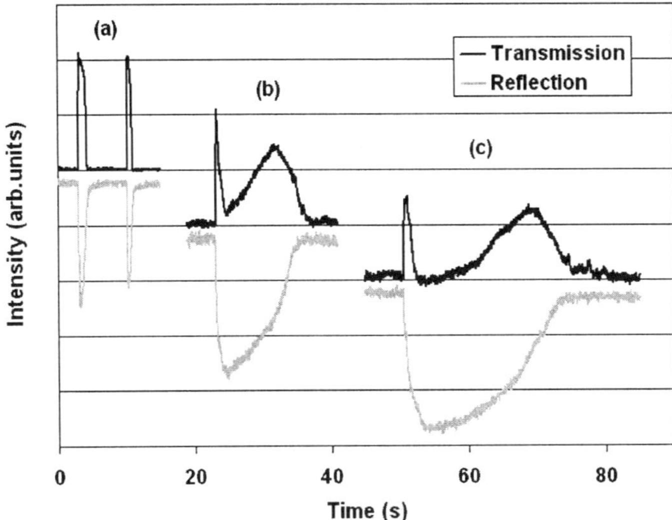

Figure 1. Sensing of condensation from respiration for (a) short, (b) medium, and (c) long breaths, as measured with the silicon dew point sensor. The duration of exhalation was approximately 0.5 s, 2 s and 5 s, respectively. Arrival of the humid air corresponds to the leading edge of the spike in the SPR transmission measurements. Two separate breaths are shown in (a).

Instead, the transmission behavior appears to reflect the dynamics of the water droplet growth. It is known that condensation initially results in a thin, continuous sheet of water coating the surface. If condensation continues (longer breaths), the water nucleates into droplets, thereby reducing the surface area coverage and allowing plasmon excitation once again. This process is gradually reversed as the droplets shrink to a critical size during evaporation. This process is compatible with previous descriptions in terms of the free energy of the droplet, given by the sum of a surface term (area times surface tension) and a volume term (volume times the negative bulk free energy change)[5].

In order to operate this apparatus as a dew point sensor, it will be necessary to cool the silicon wafer at a constant, controlled rate. Although we have not yet achieved satisfactory temperature control, these initial results confirm that the SPR method is compatible with silicon structures and that the technique is extremely sensitive to the dynamics of the surface coating process.

3. Fiber-optic SERS sensors

When enhanced by the presence of gold or silver nanoparticles, Raman scattering provides a highly specific and sensitive sensing mechanism. However, substrates used for SERS have tended to contain a wide distribution of particle sizes, which reduces the expected particle-size dependence of the peak excitation wavelength for the LSPR[2]. Improved SERS substrates with uniform size distributions are of particular interest for the development of optical fiber SERS sensors, given the potential for minimally invasive chemical sensors with low cross-sensitivities[6].

A technique known as nanosphere lithography[7] provides a relatively accessible approach to building structures smaller than 100 nm. In this case, a metal film is deposited over a self-assembled monolayer of polymer nanospheres. Triangular metal nanoparticles are formed on the substrate through the interstices of this mask and remain on the substrate after the nanospheres are removed. Although significant technical barriers restrict the application of nanosphere lithography on the tip of an optical fiber[8], it is relatively straightforward to generate a modest SERS effect from the roughened surface of the metal-coated nanospheres.

An example of this approach is shown in Fig. 2, where SERS spectra of thiophenol are presented. In this experiment, the tip of a 300 μm diameter fiber was partially coated with a monolayer of 190 nm polystyrene spheres (Interfacial Dynamics Corp.), followed by a 50 nm layer of gold. The tip was then immersed in a 10 mM solution of thiophenol in ethanol for 10 minutes, before being rinsed in ethanol. The thiophenol provides a convenient test system, as it forms relatively stable monolayers on gold surfaces.

The thiophenol spectrum was readily observed when measured directly off the tip (upper curve in Fig. 2). The main SERS peaks (indicated by the bars) can still be distinguished when measured through the 10 cm long fiber (lower curve in Fig. 2). The spectrum taken through the fiber has been partially background corrected, but now contains additional spectral features associated with the glass fiber. These results suggest that additional optimization of the SERS coating is required to raise the signal level above that of the fiber background.

Subsequent AFM measurements revealed that a surfactant residue was blocking the gaps between the nanospheres. This restricts the formation of isolated nanoparticles and reduces the SERS enhancement. More recent results have shown that improved SERS performance can be obtained using a low surfactant dispersion of 350 nm PMMA nanospheres (Soken Chemical and Engineering Co.). After drying, the reduced surfactant residue allows nanoparticle formation through the interstices of the spheres. We have also confirmed that silver films provide a stronger SERS enhancement, as found by other workers[6]. These results suggest that significantly improved performance can be obtained from SERS-structures on the tip of an optical fiber.

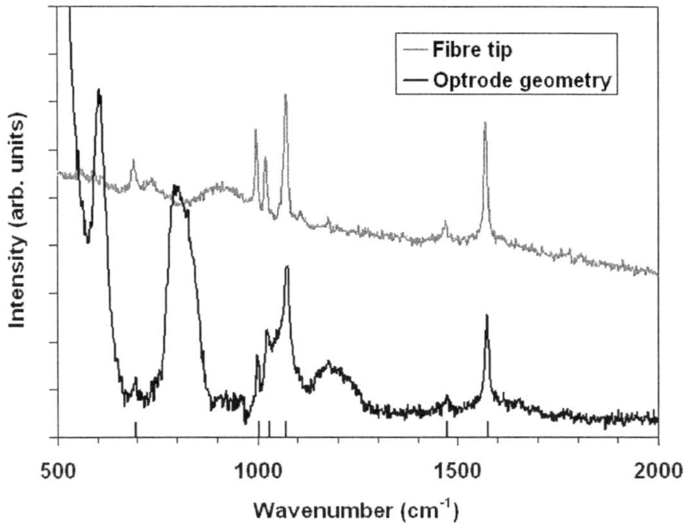

Figure 2. SERS spectra of thiophenol. This was taken from a gold film over nanospheres on the tip of an optical fiber. The spectrum was measured directly off the fiber tip (upper curve) and then through the optical fiber (lower curve).

4. Conclusion

The results presented here demonstrate that there is great potential to produce miniaturized plasmonic sensors on silicon and fiber optic platforms. In doing so, it will be possible to take advantage of the powerful materials handling and manufacturing technologies that have been developed for silicon and optical fiber devices, thereby reducing costs and improving reliability.

References

1. R. Weisendanger, *Sensors: A Comprehensive Survey*, Vol. 8, eds. W. Göpel, J. Hesse and J.N. Zemel, VCH, Weinheim, 337 (1995).
2. M. Moskovits, *Rev. Mod. Phys.* **57**, 783 (1985).
3. R.G. Wylie *et al.*, *Humidity and Moisture*, Vol. 1, eds. A. Wexler and R.E. Ruskin, Reinhold, New York, 125 (1965).
4. E. Kretschmann and H. Raether, *Z. Naturforsch.* **23A**, 2135 (1968).
5. J.P. Hirth and G.M. Pound, *Condensation and Evaporation, Nucleation and Growth Kinetics*, Pergamon Press, Oxford (1963).
6. C. Viets and W. Hill, *Int. J. Vib. Spect.* [www.ijvs.com] **4**, 8 (2000).
7. C.L. Haynes and R.P. Van Duyne, *J. Phys. Chem.* **105**, 5599 (2001).
8. P.R. Stoddart, P.A. White and A. Mazzolini, *Smart Sensors II*, Proc. SPIE Vol. 4934, eds. A.R. Wilson and V.V. Varadan, 61 (2002).

SYNTHESIS AND CHARACTERIZATION OF NEARLY MONDISPERSE CDSE QUANTUM DOTS AT LOWER TEMPERATURE

*RONG HE, HONGCHEN GU**

*Engineering Research Center of Nano Science and Technology,
Shanghai JiaoTong University, 1954 Huashan Road, Shanghai, 200030, P. R. China*

A one-pot synthesis of high-quality quantum dots of CdSe was developed using cadmium acetate as cadmium precursor. This method was proposed as an alternative to the classical organometallic route based on highly toxic, expensive, and pyroforic dimethylcadmium. The structures of the CdSe QDs were determined by X-ray powder diffraction and transmission electron micrograph. UV-Vis spectroscopy and photo luminescent spectroscopy were used to follow the reaction process and to characterize the optical properties of the resultant CdSe QDs. The obtained QDs exhibited a narrow PL band with reproducible room-temperature quantum yields as high as 30%. The emission colour is tuneable from yellow to red with increasing diameter of CdSe nanocrystals. This green-chemistry approach may also be used for time-resolved, in situ study of crystallization of the quantum dots. The scheme is reproducible and simple and thus can be readily scaled up for industrial production. General aspects of the nanocrystal nucleation and growth were also discussed. The properties of the CdSe QDs modified with organic ligands such as PVP, phosphatide, phosphatidyl choline and 2-hydroxyl-4-*n*-octoxide-diphenyl ketone were also investigated in this report.

1. Introduction

Colloidal semiconductor nanocrystals, also known as quantum dots (QDs), are of tremendous fundamental and technical interest due to their applications as light-emitting devices, low-threshold lasers, and biological labels. Owing to their size-dependent photoluminescence (PL) tuneable across the visible spectrum, CdSe nanocrystals have become the most extensively investigated QDs. Organometallic routes using dimethyl cadmium ($Cd(CH_3)_2$) as the cadmium precursor has been well developed [1,2]. The toxic, pyrophoric, expensive precursor, high reaction temperature and restricted equipments limited this method scaled up for industrial production. In a recent development, Peng [2] reported the use of CdO complexed to hexylphosphonic acid (HPA) or tetradecylphosphonic acid (TDPA), as an alternative to $(CH_3)_2Cd$, in TOPO syntheses of CdE (E = S, Se, Te). In this paper, we will study the possibility of synthesis of CdSe using $CdAc_2$ as precursor with different organic ligands at lower temperature under mild conditions.

---

* Corresponding author, Fax: +86-21-62804389, Tel: +86-21-62933731;
  E-mail: hcgu@sjtu.edu.cn

2. Experimental section

In a typical procedure, 0.001mol cadmium acetate (CdAc$_2$) and 1ml of oleic acid or other organic ligands were dissolved in 4ml of phenyl ether. The reaction mixture was heated to proposed temperature for 1 h under stirring and under a continuous flow of nitrogen and then cooled. When the temperature fell below 60℃, 2ml of 1M trioctylphosphine selenide (TOPSe) was added to form the precursor solution. Then the precursor solution was injected into a flask containing 10ml phenyl ether. The reaction temperature was varied from 110 to 170 ℃. Aliquots of the reaction mixture were taken at 1min intervals during the first 10min, and then at longer intervals. The sampling aliquots were quenched in cold chloroform in order to stop further growth of the particles.

The ultraviolet-visible spectra were measured in 1cm quartz cuvettes in air with a Unicam-UV 300 spectrophotometer. Chloroform was used as the reference. Room-temperature photoluminescence spectra were recorded on a Hitachi F-2500 fluorescence spectrophotometer. Samples were prepared by dispersing washed CdSe nanocrystallites in chloroform. Estimates of quantum yield were obtained by comparing the integrated emission from Rhodamine 6G in ethanol and that of CdSe nanocrystallites dispersed in chloroform. Optical densities of all solutions were below 0.10 at the excitation wavelengths used. Transmission electron microscopic analysis was performed with a Hitachi 600 microscope operating at 120kV. Samples were prepared by drying a drop of the colloids on a TEM grid and the sample allowed to dry completely at room temperature. Approximately 200 nanoparticles were measured manually for size distribution. The X-ray diffraction patterns were recorded on a Shimadzu XD-3A X-ray diffractometer using Cu K$_\alpha$ radiation (λ =0.1542nm) operated at 50 kV and 100mA.

3. Results and Discussion

3.1 Optical Characterization

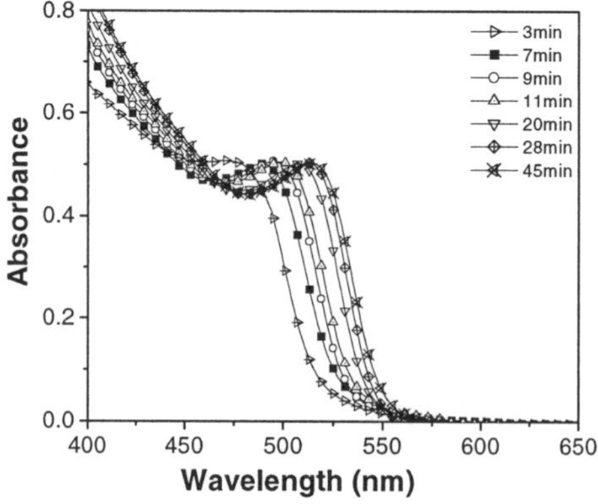

Figure 1. Temporal evolution of absorption spectra of CdSe QDs

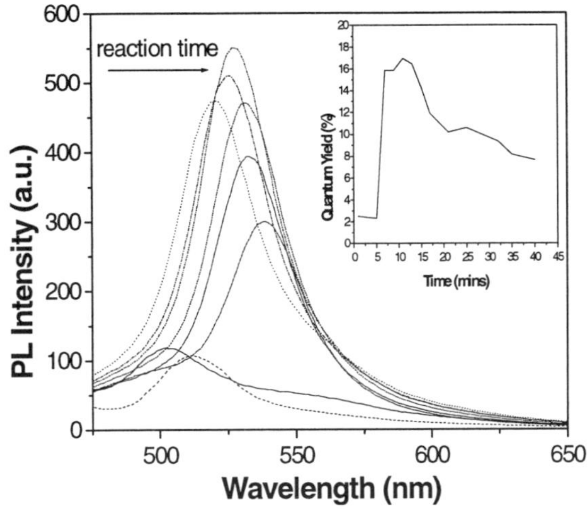

Figure 2. Temporal evolution of PL and PL quantum efficiency (inset)
spectra of as-prepared CdSe nanocrystals

The UV-Vis (Fig. 1) and PL (Fig. 2) spectra were used to follow the reaction process and to characterize the optical properties. The sharp absorption features suggest highly monodisperse samples. High quantum yields and narrow emission line widths indicate growth of crystallites with few electronic defect sites. Both the absorption and the PL spectra of the QDs shifted to longer wavelengths as the particle size increased as a consequence of the quantum confinement effect. It is clearly that the QY rapidly reached a maximum and decreased afterward.

3.2 Influence of the Growth Temperature

Table 1 The absorption and emission peaks at different growth temperature

Growth Temperature (℃)	Absorption Peak (nm)	Emission Peak (nm)
170	515	540
140	501	526
110	476	500

The absorption and PL spectra showed that the final size of CdSe QDs determined by the reaction temperature (table 1). At higher temperatures larger particles are obtained. The formation of the larger particles at higher temperature can be explained by taking into account the kinetics of the particle formation. A higher rate of attaching and detaching at higher temperatures results in a faster growth rate and thus larger particles size.

3.3 Structural Characterization

Typical XRD pattern of the prepared CdSe QDs is showed in Fig. 3a. The three distinct diffraction peaks match exactly the (111), (220) and (311) crystalline planes of cubic CdSe, indicated the formation of CdSe. The TEM image in Fig. 3b showed nearly monodisperse CdSe nanocrystals.

62

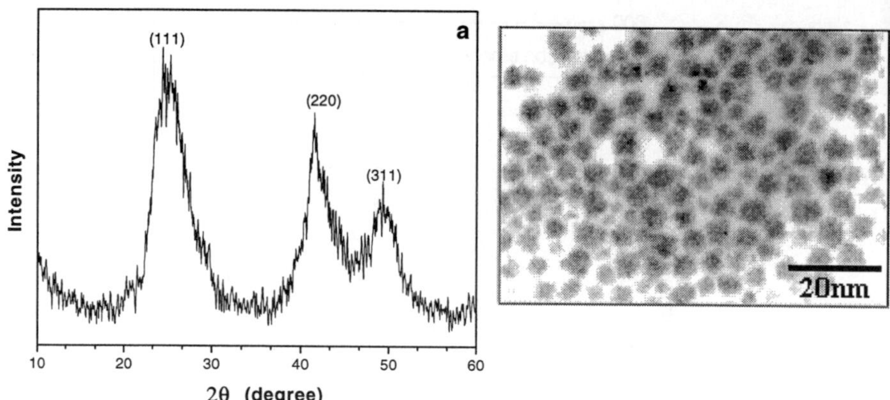

Figure 3. Typical XRD pattern and TEM image of as-prepared CdSe QDs

3.4 Influence of the Ligands

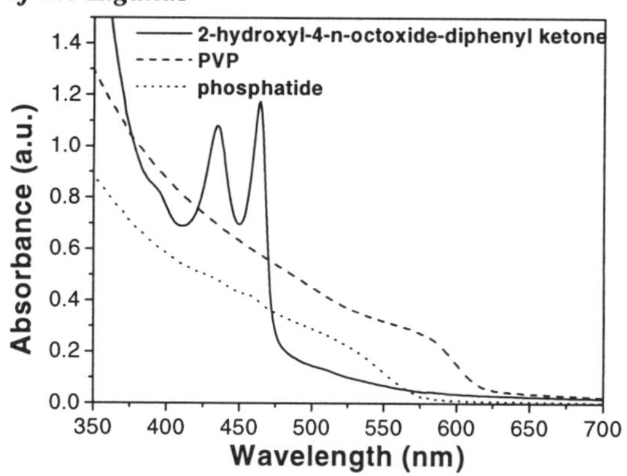

Figure 4. UV-Vis spectra of as-prepared CdSe QDs with different ligands

It is well known that the choice of ligands is crucial for the formation of nm-sized semiconductor crystals. Fig. 4 showed that the peak position and the shape of the absorption spectra were clearly distinguishing when different ligands were used. The exact interactions between nanoparticles and organic ligands will be studied in detail.

Main References

1. C. B. Murray, D. J. Norris, and G. Bawendi, *J. Am. Chem. Soc.* **115**, 8706, (1993).
2. C. M. Dognega, S. G. Hickey, S. F. Wuister, D. Vanmaekelbergh, and A. Meijerink, *J. Phys. Chem. B* **107**, 489, (2003).
3. Z. A. Peng and X. G. Peng, *J. Am. Chem. Soc.* **123**, 183, (2001).

UNCONVENTIONAL FERROMAGNETISM IN ALL-CARBON CLUSTER-ASSEMBLED NANOFOAM

A. V. RODE, E. G. GAMALY, A. G. CHRISTY, J. G. FITZ GERALD, S. T. HYDE,
R. G. ELLIMAN, B. LUTHER-DAVIES

The Australian National University, Canberra, ACT 0200, Australia

J. ANDROULAKIS, J. GIAPINTZAKIS

Foundation for Research and Technology-Hellas,
Institute of Electronic Structure and Lasers, Heraklion, Crete, Greece

We report production of nanostructured carbon foam by a high-repetition-rate, high-power laser ablation of glassy carbon in Ar atmosphere. A combination of characterization techniques revealed that the system contains both sp^2 and sp^3 bonded carbon atoms. The material is a novel form of carbon in which graphite-like sheets fill space at very low density due to strong hyperbolic curvature, as proposed for "schwarzite". The foam exhibits ferromagnetic-like behaviour up to 90 K, with a narrow hysteresis curve and a high saturation magnetization. Such magnetic properties are very unusual for a carbon allotrope. Detailed analysis excludes impurities as the origin of the magnetic signal. We postulate that localized unpaired spins occur because of topological and bonding defects associated with the sheet curvature, and that these spins are stabilized due to the steric protection offered by the convoluted sheets.

1. Cluster-Assembled Carbon Nanofoam

The discovery of new nanostructured carbon phases, including fullerenes and nanotubes, has opened a new era in materials science. Minute changes in the spatial arrangement of carbon atoms can profoundly alter the electronic properties of these systems from a semiconductor to a metal or superconductor. One could expect that structural rearrangements might also significantly change the magnetic properties of all–carbon allotropes from their known diamagnetic behaviour. Only recently, a weakly magnetic all–carbon structure has been reported, formed under high-pressure and high-temperature conditions, consisting of rhombohedrally polymerized C_{60} molecules.[1] Here we report the occurrence of ferromagnetism with a very high density of unpaired spins up to 1.51×10^{20} per gram, in a novel semiconducting nanostructured carbon foam with a low mass density. The nanofoam is shown to be a soft ferromagnetic semiconductor with a narrow hysteresis curve and a Curie temperature exceeding 90 K. Our results suggest that unpaired spins are introduced by sterically protected carbon radicals, which are immobilised in the non-alternant aromatic system of carbon sheets with negative Gaussian curvature.[2, 3]

1.1. *Synthesis*

We have recently synthesized a new form of carbon, a cluster assembled carbon nano-foam, by high-repetition-rate laser ablation.[3-5] The basic principle behind this new laser ablation regime suggested the use of short, picoseconds (10^{-12} s) laser pulses delivered on a target with a repetition rate of up to several tens of megahertz. In such conditions, the ablation process enters the steady-state regime where the continuous influx of atoms from the ablation surface matches the loss due to diffusion and a steady plume is created.[5]

The low-density cluster-assembled carbon nanofoam was produced by high-repetition-rate laser ablation of an ultra-pure glassy carbon target in a vacuum chamber with the base vacuum ~5×10^{-7} Torr, filled with high-purity (99.995%) Ar gas to ~100 Torr, inside a 2" cylinder made of fused silica (SiO_2). The carbon vapor temperature in the laser plume, where the formation process takes place, is in the range 1-10 eV (~10,000 - 100,000 K), i.e. the formation takes place in partly ionized plasma. The high-repetition-rate laser ablation creates an almost continuous inflow of hot carbon atoms and ions with an average temperature of ~2 eV into the experimental chamber. This vapor heats the ambient Ar gas and increases the partial density of carbon atoms in the chamber. The process of formation of carbonaceous clusters begins when the carbon density reaches the threshold density, at which the probability of collisions between carbon atoms becomes sufficiently high.[5] Diffusion-limited aggregation during the subsequent fast quench resulted in a unique, fractal all-carbon system.

1.2. *Characterisation*

In order to check the reproducibility of our results we have produced several samples, which were characterized by means of scanning electron microscopy (SEM), transmission electron microscopy (TEM), high resolution transmission electron microscopy (HRTEM), electron energy loss spectroscopy (EELS), Rutherford back-scattering (RBS), and trace elemental analysis by induction-coupled plasma mass spectrometry (ICP-MS) of acid extracts. The produced carbon nano-foam possesses a fractal-like structure consisting of carbon clusters with an average diameter of 6-9 nm randomly interconnected into a web-like foam (Fig.1). This material exhibits some remarkable physical properties like the lowest measured gravimetric density (~2 mg/cm^3) ever reported for a solid, and a large surface area (comparable to zeolites) of 300-400 m^2/g. Our structural studies revealed the presence of hyperbolic "schwarzite" structures.[4] Schwarzites are anticlastic (saddle-shaped) warped graphite-like layers, analogous to the synclastic sheets in fullerenes.[6] HRTEM images suggest periodic structures within the individual clusters, with a period of ~5.6Å (Fig. 2).[3-5]

Fig. 1. Transmission electron micrograph (TEM) of the carbon nanofoam (left). One can see that the foam is assembled from 6 nm clusters. Scanning electron micrograph (SEM) at right shows the web-like structure at lower magnification.

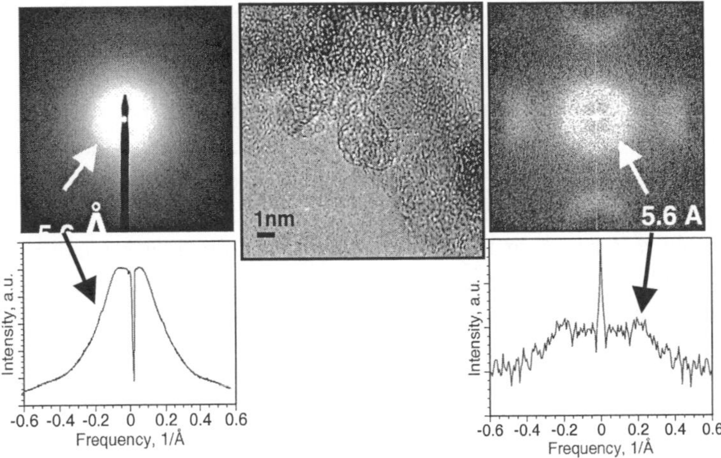

Fig. 2. High-resolution TEM image of the carbon nanofoam (central image), with an individual 6-nm cluster clearly seen in the centre. Fourier transforms (right) and electron diffraction patterns (left) indicate a 'repeat' spacing with characteristic length of 5.6±0.4 Å typical for 'schwarzites'.

The most salient property of the foam is its unusual magnetic behaviour. Unlike other carbon allotropes, known to be diamagnetic, the freshly produced foam is initially attracted strongly to a permanent 4 kOe $Nd_2Fe_{14}B$ magnet at room temperature, thus demonstrating the existence of a substantial permanent magnetic moment. The initial room-temperature magnetization was found to disappear within few hours after production due to relaxation processes. In the following, we report the magnetic behaviour of the equilibrated system.

1.3. *Chemical impurities analysis*

The possibility of magnetic contamination, during synthesis, of the nanofoam samples was thoroughly examined. The impurity content in the foam was determined from 2-MeV He^+ ion RBS measurements, and independently by mass spectroscopy analysis of acid extracts from the foam. Both methods show comparable low impurity contents. The sum of all the impurity elements analyzed in the foam of this study was 415 - 465 ppm, with 30 ppm of *Ni* and *Fe* < 80 ppm. The dominant impurity elements in the samples were Al, Cu, Zn, Pb, Ni, and Fe. The elemental makeup, in combination with considerable run-to-run variability of impurity content are compatible with the hypothesis that the impurities are introduced randomly as metallic dust particles (primarily aluminum, brass, solder and stainless steel) originating in the apparatus. We eliminate impurities as significant contributors to the measured magnetic properties.

2. Magnetization measurements

All magnetization measurements were performed in a commercial extraction magnetometer (Maglab Exa 2000) by Oxford Instruments in the temperature range $1.8 \leq T \leq 300$ K and in applied magnetic fields up to 70 kOe. Reproducibility of the magnetic properties of the foam is demonstrated by the comparable magnetization (0.36 - 0.8 emu/g) measured in six independently synthesized batches of foam 15 days after synthesis. The magnetization of the foam in this study (0.42 emu/g at 60 days and 12 months after the synthesis) is evidently typical for the material.

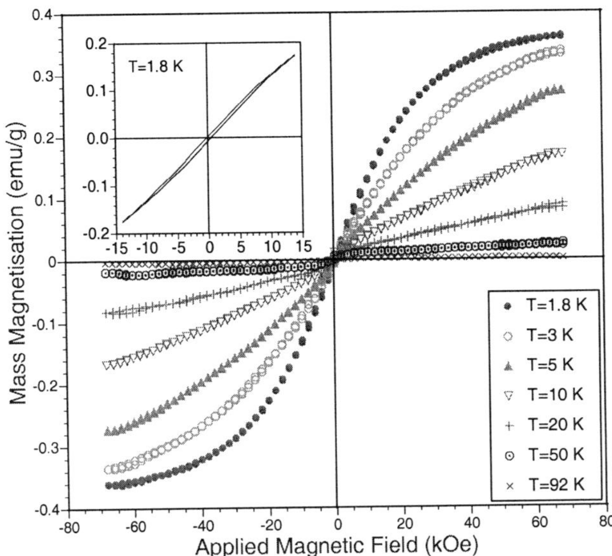

Fig. 3. Mass magnetization of the foam as a function of the applied magnetic field, $M(H)$, at several temperatures from 1.8 to 92 K. All data are corrected for the diamagnetic contribution of the gelatine sample holder. Inset: $M(H)$ hysteresis loop at T = 1.8 K exhibiting a coercive force H_c = 420 Oe.

The observed signal is positive and apparently paramagnetic (*PM*). However, the expected *PM* $1/T$-dependence was not observed, which is indicative of a non Curie-Weiss system. Hence, the magnetization isotherms at low temperatures were investigated. Figure 3 illustrates the mass magnetization of the foam as a function of the applied magnetic field at several temperatures from 1.8K to 92K. All data have been corrected for the diamagnetic contribution of the gelatine sample holder that was again measured separately under the same conditions. The measured signal is positive and the curves show *PM*-like behaviour. Nevertheless, we observe a slight hysteresis with a well-defined coercive force at low temperatures (see inset of Fig. 3) as expected for a ferromagnet (*FM*).

$M(H)$ data taken at $T = 1.8$K, were plotted against $1/H$ in order to obtain the saturation magnetization of 0.42 emu/g after extrapolation to infinite field. This is equivalent to a saturation moment value of 9.0×10^{-4} μ_B per carbon atom (μ_B is the Bohr magneton equal to 9.27×10^{-21} erg/G). Assuming that our spin system is *FM*-like rather than *PM* (i.e., $1\mu_B$ per unpaired spin), we estimate that this value corresponds to about 1 unpaired spin per 1000 carbon atoms, a fact which suggests that several unpaired spins are located in each of the nanometre-scale spheroidal clusters with $\sim 10^4$ C-atoms/cluster that constitute the foam,[4] and is in order-of-magnitude agreement with low temperature ESR measurements that show a large concentration of unpaired spins 1.8×10^{20}/g (3.6 unpaired spins per 1000 carbon atoms).

3.   Discussion of magnetization data

We have observed a strong positive magnetization signal in a new all-carbon structure, which seems to have features of both a *PM* and a *FM*. It could be hypothesized that our observed sample behaviour combines a *FM* signal from chemical impurities (the $3d$ elements) and *PM* from the new carbon phase (foam). However, the experimental evidence does not support this possibility. First, if we assume an absolute worst-case scenario in which all the Fe, Co and Ni were in their ferromagnetic elemental forms, the impurity magnetization would be at most 0.09 emu/g in our sample, which is only 20% of the measured total. Second, the observed response of sample magnetism to temperature and applied magnetic field is not what would be expected from transition metal bearing ferromagnetic impurities. And finally, the carbon nano-foam was found to exhibit a strong magnetic relaxation in time, achieving magnetic equilibrium several weeks after production. That is incompatible with conventional *FM* impurities. Thus, the observed magnetic behaviour is an intrinsic property of the nano-foam.

The magnetic behaviour exhibited by the new phase of carbon is extremely unusual. It differs from the weak positive magnetization found at very low temperatures in single-walled nanohorns and activated carbon fibers, mentioned in the introduction. In these two cases, the occurrence of magnetism has been associated with exposed graphitic edges. Our observations point to a different, unique origin for the magnetism in the foam. We believe that the remarkable magnetic properties of the foam, unexpected for an all-carbon material, are an intrinsic consequence of its equally remarkable nanostructure.

HRTEM shows convoluted graphite-like layers inside the nano-spherular building blocks of the foam, the contrast being consistent with hyperbolic schwarzite curvature of the sheets.[4] This requires rings of 7 or more carbons interspersed with normal graphitic 6-rings.[7] The sheet curvature localizes unpaired spins by breaking the continuity of the delocalized π-electron clouds of graphite, and tight curvature of the sheets provides a mechanism for sterically protecting the unpaired spins which would otherwise be too chemically reactive to persist.[3]

A possible mechanism for magnetic moment generation would be a simple indirect exchange interaction through conduction electrons located on the hexagons. This, however, has to be in agreement with electrical resistivity measurements that show a semiconducting behaviour with a band gap of 0.5-0.7

eV. A plausible scenario would be that the magnetism of our nano-foam is an effect that actually occurs in nano-sized metallic (sp^2) segments of the structure that are isolated by non-conducting (sp^3) regions, and hence do not contribute to the overall conductivity of the sample.

We have observed unique magnetic behaviour in an all-carbon nano-structure, the unusual structure of which provides a plausible mechanism for generation of strong magnetism. Our data leads us to reject ferromagnetic impurities as the origin of the observed magnetism. Combining our experimental results, M vs H/T scaling, magnitude and temperature dependence of the moment of ferromagnetic impurities, and strong time-dependent magnetization relaxation, lead us to safely conclude that the observed behaviour is an intrinsic property of the foam itself. The behaviour does not fit into the categories of conventional paramagnetism or super paramagnetism (data do not fit the Brillouin function with $S = 1/2$, and M vs H/T curves do not collapse onto one another). We have observed small hysteresis and remnant magnetization in the $M(H)$ curve of our foam which are usually observed in organic ferromagnets[8] and hence do not exclude the case of weak soft ferromagnetism. Nevertheless, we have no clear signs of an ordering temperature and most importantly the structure of our foam suggests that the clusters of the foam are too small to sustain a permanent magnetic moment as in normal ferromagnetic materials.

This new form of carbon clearly warrants further theoretical and experimental investigations.

References

1. T. L. Makarova, B. Sundqvist, R. Höhne, P. Esquinazi, Y. Kopelevich, P. Scharff, V. A. Davydov, L. S. Kashevarova, A. V. Rakhmanina, *Nature* **413**, 716-718 (2001).
2. N. Park, M. Yoon, S. Berber, J. Ihm, E. Osawa, D. Tomanek, *Magnetism in all-carbon structures with negative Gaussian Curvature*, accepted for publication in Phys Rev Lett. (October 2003).
3. A. V. Rode, R. G. Elliman, E. G. Gamaly, A. I. Veinger, A. G. Christy, S. T. Hyde, B. Luther-Davies, *Appl. Surf. Sci.*, **197-198**, 644-649 (2002).
4. A. V. Rode, S. T. Hyde, E. G. Gamaly, R. G. Elliman, D. R. McKenzie, S. Bulcock, *Appl. Phys.* **A69**, S755-S758 (1999).
5. A. V. Rode, E. G. Gamaly, B. Luther-Davies, *Appl. Phys.* **A70**, 135-144 (2000).
6. D. Vanderbilt and J. Tersoff, *Phys Rev Lett.*, **68** 511-513 (1991).
7. S. T. Hyde, M. O'Keeffe, *Phil. Trans. Roy. Soc. Lond. A* **354**, 1999 (1996).
8. A. Mrzel, A. Omerzu, P. Umek, D. Mihailovic, Z. Jaglicic, Z. Trontelj, *Chem. Phys. Lett.* **298**, 329 (1998).

THE APPLICATIONS OF FOCUSED ION BEAM TECHNOLOGY TO NANOSCALE MATERIALS

J.M. CAIRNEY and P.R MUNROE
Electron Microscope Unit
University of New South Wales
SYDNEY NSW 2052
Australia

Focused ion beams have been used for some time for defect analysis and integrated circuit (IC) repair in the semiconductor industry. More recently they have become a valuable tool in materials characterization. The development of dual-beam technology, combined with *in-situ* gas control, a range of possible source materials and lithographic software mean the instrument is well placed to be used in the fabrication, as well as observation, of materials on the nano-scale.

1. Introduction

Electron microscopy is now an extremely mature materials characterization technique. The use of high-energy electron beams yields a wealth of data on the structural, chemical and crystallographic nature of materials. More recently, high energy, focused ion beams (FIBs) have found applications in the analysis of materials. The use of, most usually, gallium ions, allows materials to be sectioned with very high spatial precision. Further, the secondary electrons and ions that are emitted from the specimen allow high-resolution images to be obtained. The development of the 'dual beam' FIB, which combines an ion column with a high resolution electron column, along with accessories such as lithographic software, gas injection systems and nano-manipulators, allow this technology to be used to fabricate materials as well as characterize them.

In this paper the basic principles of focused ion beam milling will be described briefly. Subsequently, a number of examples of the range of materials analysis and materials fabrication that can be performed will be described.

2. Focused Ion Beam Technology

At face value, a FIB system resembles a scanning electron microscope (SEM). A beam of high-energy particles is scanned over the surface of the specimen. Secondary electrons emitted from the specimen are detected and used to form images. However, there are some important differences between a SEM and a FIB. Most commercial FIB systems use a high-energy (typically 30 keV) gallium ion beam. The gallium source itself is a field emitter and the arrangement of a high field, called the extraction voltage, around the tip of the source allows a fine 5-10 nm, intense beam of ions to be generated. Unlike, SEMs, FIBs use electrostatic, rather than magnetic, fields to focus the beam.

Gallium ions are many orders of magnitude heavier than electrons. When they impact on the specimen they may cause atoms in the target (specimen) to be sputtered. Some of the sputtered atoms are positive ions and these can be detected to generate a secondary ion image. The interaction leads to the ejection of secondary electrons that can also be detected. The beam current incident on the specimen is controlled by a beam-limiting aperture. If a large, ~ 100 microns, diameter aperture is inserted, the beam current is on the nA scale and rapid sputtering takes pace. Sections several microns square can be cut within a few minutes. If a smaller beam-limiting aperture is inserted then the beam current is restricted to a few pA and the signal generated can be used to form images. Of course, even at such beam currents some sputtering will take place.

3. Applications of FIB Technology

The most conventional application of the FIB is the preparation and examination of cross-sections. Sections up to about 10 microns in depth can be prepared in a few minutes. If the specimen is tilted the cross-section can be viewed and analyzed. Figure 1 shows a cross-section through a nano-indented TiN film [1]. The FIB is able to prepare sections precisely through the indentation to allow observation of the sub-surface deformation mechanisms. A second common application is the preparation of transmission electron microscope sections. Again, the focused ion beam can be used to prepare electron transparent sections with highly localised precision. The FIB is ideal for preparing sections through specimens which have undergone surface deformation, oxidation or are in thin film form, for example, highly brittle quasi-crystal coatings deposited on FeAl-based substrates [2].

The above applications are reasonably well-known and understood. However, the FIB is now being used not only in the analysis of materials, but in the fabrication of materials. There are a number of ways in which the FIB can be used in fabrication. Firstly, the energetic ion beam can be used for implantation. The implantation of gallium can be used to change the behaviour of dielectric materials. For example implantation of gallium into SiC has been shown to change its optical properties [3]. Furthermore, a range of source materials are being developed which will allow the implantation of species other than gallium [4]. Species such as germanium, cobalt and silicon may be implanted into substrates. The nature of the focused ion beam allows this to be done without lithographic masking.

It is well established that gas injection systems can be used in the FIB. These introduce gaseous species into the vicinity of the specimen, which under the energetic action of the ion beam react and lead to either the deposition of species or locally enhanced sputtering rates. A number of materials can be readily deposited, these include both conductors such as platinum or tungsten or dielectrics such as silicon oxide. Such *in-situ* deposition can be used in a number of applications such as the joining of components, re-wiring of integrated circuits or the protection of surfaces prior to milling.

The FIB is also increasingly used in applications in lithography and nano-machining. The gallium ion beam can cut trenches and sections with a precision of about 10 nm. This readily allows the generation of two dimensional structures that can be used, for example, as gratings for optical or x-ray diffraction [5,6]. Figure 2 shows such a grating prepared for super-resolution optical microscopes from chromium-coated glass slides. The current generation of focused ion beam instruments allow the introduction of nano-manipulators into the specimen chamber [7]. This allows the specimen to be geometrically manipulated permitting series of either sections or depositions to be made in sequences which allow the fabrication of more complex devices in three dimensions [8].

4. Summary

The FIB is an invaluable tool in materials characterization. The current generation of FIB's which may include nano-manipulators, non-gallium sources and lithographic software also allow the fabrication of materials on a nano-scale.

References

1. L.W. Ma, J.M. Cairney, M.J. Hoffman and P.R. Munroe, *Applied Surface Science*, in submission, (2003)
2. J.M. Cairney, P.R. Munroe and D.J. Sordelet, *J. Microscopy,* **201**, 201 (2001).
3. T. Tsvetkova, O. Angelov, M. Sendova-Vassileva, D. Dimova-Malinovska, L. Bischoff, G. J. Adriaenssens, W. Grudzinski and J. Zuk, *Vacuum.* **70**, 467 (2003).
4. T. Ganetsos, C. Aidinis, L. Bischoff, G.L.R. Mair, J. Teichert, D. Panknin and I. Papadopoulos, *Journal Of Physics D-Applied Physics.* **34**, 839 (2001).
5. T. Thio, H.J. Lezec, T.W. Ebbesen, K.M. Pellerin, G.D. Lewen, A. Nahata and R.A Linke, *Nanotechnology*, **13**, 429 (2002).
6. R. Heintzmann, V. Sarafis, P.R. Munroe, J. Nailon, Q.S. Hanley, T.M. Jovin, *Micron*, **34**, 293 (2003).
7. R.M. Langford, *Micron*, in press, (2003).
8. S. Reyntjens and R. Puer, *J. Micromech. Microeng.* **10,** 181 (2000).

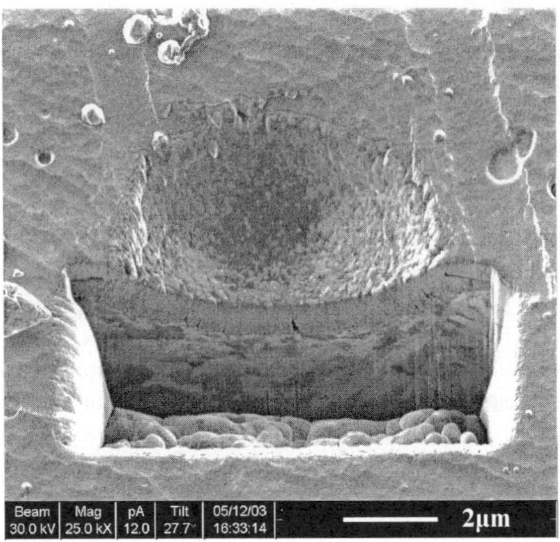

Figure 1: FIB section milled through indented TiN coating on a steel substrate

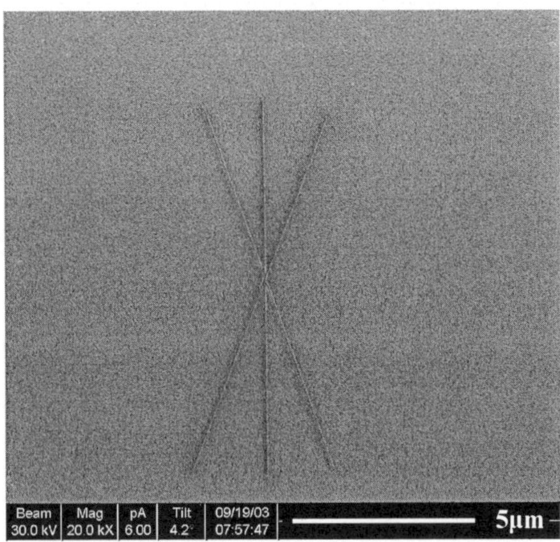

Figure 2: FIB milled sections in Cr coating on glass for application as an optical grating

LOW-TEMPERATURE FORMATION OF SiGe NANO-CRYSTALS BY METAL-INDUCED LATERAL CRYSTALLIZATION

M.MIYAO, H.KANNO, T.SADOH, AND A.KENJO

Depart. of Electronics, Kyushu University, 6-10-1 Hakozaki
Fukuoka 812-8581, Japan

Low-temperature ($<550^{\circ}$C) formation of SiGe nanocrystals by metal-induced crystallization of a-$Si_{1-x}Ge_x$ films on SiO_2 layers has been investigated. Plane growth was observed for Ge fractions below 20 %. The grain size increased with increasing Ge fraction, and strain-free poly-$Si_{0.8}Ge_{0.2}$ films with large grains (15 μm) were obtained. On the other hand, dendrite growth was dominant for intermediate Ge fractions with 40-60 %. Directions and widths of dendrites became straight and narrow with decreasing annealing temperature. As a result, very sharp needlelike nanocrystals (width: 50 nm, length: 10 μm) were obtained by optimizing the growth conditions (x: 0.4, annealing: 450 $^{\circ}$C, 20 h). The SiGe nanocrystals on SiO_2 layers can be useful for the advanced system-in-displays and novel one-dimensional wires.

1. Introduction

Low temperature ($<550^{\circ}$C) formation of SiGe heterostructures on insulating films has been expected to realize advanced three-dimensional ULSI and system in display. To achieve this, recrystallization processes of amorphous SiGe (a-SiGe) on SiO_2 have been widely investigated. However, only poly-SiGe with small grains (<0.1μm) was obtained by solid-phase recrystallization. Melt-grown process such as laser annealing achieved poly- SiGe with large grains (~5μm), however Ge atoms were not distributed uniformly in the films, and surface ripples with ~15 nm height were observed [1].

Recently, low temperature (~550°C) solid-phase recrystallization of a-Si was realized by using the catalytic effect of some metals [2]. This metal-induced lateral crystallization (MILC) achieved poly-Si with large grains (~10μm). In the present work, we have applied this technique to recrystallization of a-SiGe and report our findings of Ge-fraction dependence of MILC.

2. Experimental Procedures

In the experiment, a-$Si_{1-x}Ge_x$ layers (50 nm thickness, $0<x<1.0$) were deposited on SiO_2 films by using a molecular beam epitaxy system (base pressure: 5×10^{-11} Torr). Subsequently, Ni films (5 nm thickness) were evaporated on top of the a-$Si_{1-x}Ge_x$ and then patterned by using the lift-off process with photolithography.

Finally, the samples were annealed at 450-600°C in a nitrogen ambient. The crystal structures and quality of the grown layers were evaluated with Nomarski optical microscopy, scanning electron microscopy (SEM), and Raman spectroscopy.

3. Results and Discussion

Figures 1(a)-1(f) show optical micrographs of the samples with different Ge-fractions after annealing. They are classified into three groups: (i) the samples with low Ge-fractions (Figs.1 (a) and 1(b)), which crystallized with plane morphology around the Ni-pattern, (ii) those with intermediate Ge-fractions (Figs.1(c) and 1(d)), which show dendrite morphology, and (iii) those with high Ge-fractions (Figs.1 (e) and 1(f)), which scarcely crystallized.

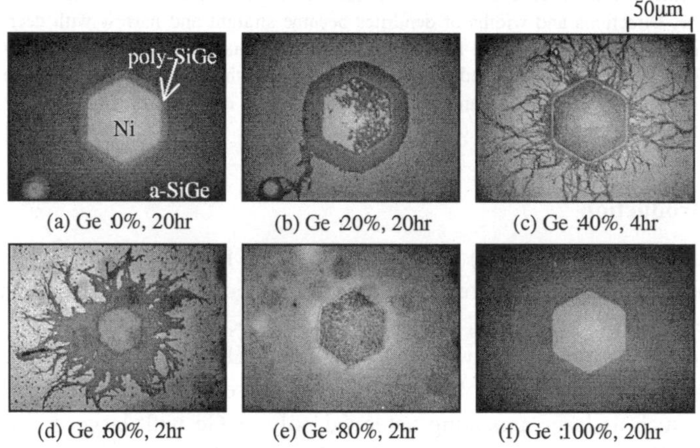

Figure 1. Nomarski optical micrographs of samples with different Ge-fractions after annealing at 550°C.

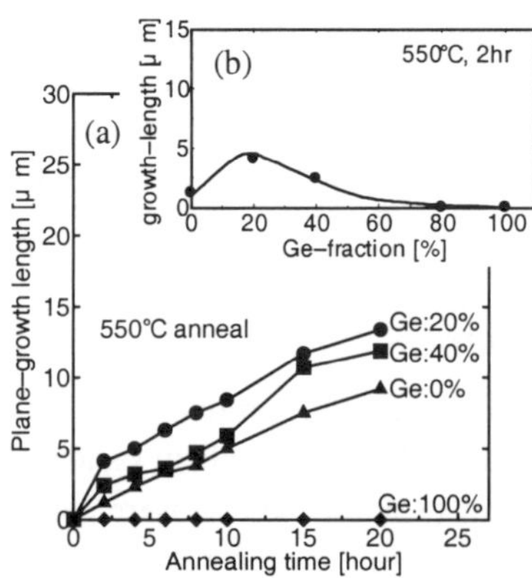

Figure 2. Lateral-length of plane-growth as a function of annealing time (a) and Ge-fraction (b).

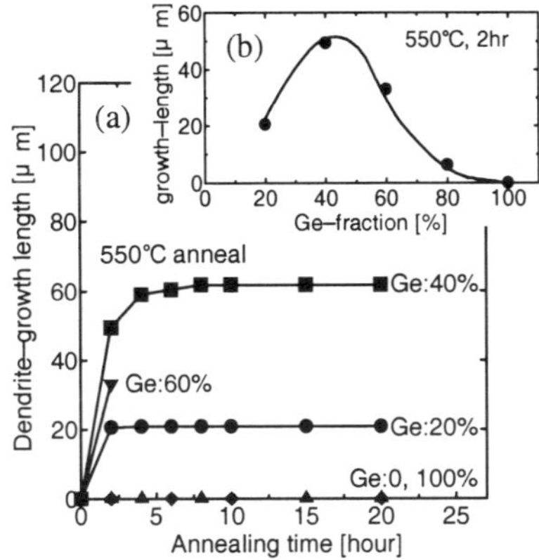

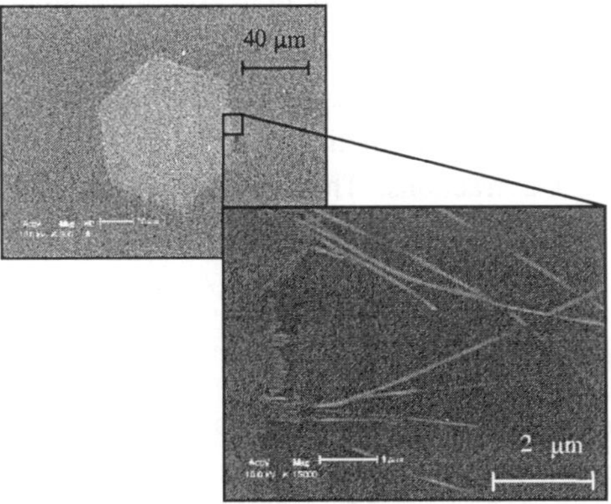

Figure 3. Lateral-length of dendrite-growth as a function of annealing time (a) and Ge-fraction (b).

Figure 4. SEM photographs for sharp dendrite-growth of $Si_{0.6}Ge_{0.4}$ after annealing at 450°C for 20 hr.

Characteristics of plane-growth in MILC-SiGe are summarized in Figs.2(a) and 2(b) as a function of annealing time and Ge fraction, respectively. Results indicate that the plane-growth progresses slowly with annealing time and reaches to 10~15μm. The growth velocity is enhanced by 50% by increasing Ge fractions from 0 to 20%, which is attributed to enhanced Ni-migration in locally strained a-SiGe films. The Raman measurements showed that the SiGe layers were completely strain free [3]. In this way, low-temperature SiGe growth by MILC method was achieved. Electrical characterizations of these crystallized regions are now underway.

Lateral-length of dendrite-growth is summarized in Figs.3(a) and 3(b) as a function of annealing time and Ge-fractions, respectively. Results indicate that dendrite-growth progresses rapidly and saturates in a very short time. Significant long growth-length of 60μm is obtained for samples with Ge-fraction of 40%.

The growth direction became straight, and the width of dendrite regions became narrower with increasing Ge fraction and decreasing annealing temperature. The mechanisms of such phenomena are not clarified yet, however, very sharp needlelike crystals (width: 50nm, length: 10μm) were obtained by optimizing the growth condition (Ge fraction: 40%, annealing: 450°C, 20hr). A typical example of the sharp dendrites is shown in Fig.4. Such needlelike crystals might be used for unique devices such as one-dimensional wires in advanced ULSI.

4.   Conclusion

Metal-induced low-temperature ($<550^{\circ}$C) recrystallization of a-SiGe on SiO_2 have been investigated. Growth velocity of MILC-SiGe can be enhanced by 50% by increasing Ge fraction from 0 to 20%, which achieved poly-SiGe with large grains (15μm). In addition, needlelike dendrite crystals were found for samples with intermediate Ge fractions. These new nanocrystalline SiGe on insulator should be used for advanced three-dimensional ULSI, system in display, and novel one-dimensional wires.

Acknowledgments

The authors are very grateful to Dr. S. Yamaguchi of C.R.L., Hitachi for his stimulating discussions. A part of this work was supported by the Grant-in-Aid for Scientific Research from the Ministry of Education, Culture, Sports, Science, and Technology of Japan.

References

1. M. Miyao, T. Sadoh, S. Yamaguchi, and S. K. Park, *Proc. 2001 Asia-Pacific Workshop on Fundamental and Application of Advanced Semiconductor Devices*, Cheju (Korea, 2001) p.115.

2. J.S.Im, *MRS Bulletin/March,* p.39 (1996).

3. H.Kanno, I.Tsunoda, A.Kenjo, T. Sadoh, and M.Miyao, *Appl. Phys. Lett.* **31**, 2148 (2003).

CATALYSIS AND CAPACITANCE OF NANOSTRUCTURED GOLD SPONGES

M.B. CORTIE

Institute for Nanoscale Technology, UTS, PO Box 123, Broadway NSW 2007, Australia

E. VAN DER LINGEN AND G. PATTRICK

Physical Metallurgy Division, Mintek, Private Bag X3015, Randburg 2125, South Africa

In the present paper we describe the catalytic properties and electrolytic double-layer capacitance of nano-structured, mesoporous gold sponges. These materials are effective catalysts for CO oxidation, and for the selective catalytic conversion of NOx. The possible application of mesoporous gold in electric double layer capacitors is premised on its high surface area, corrosion resistance and excellent electrical conductivity. The niche, if any exists, would be in high-efficiency, and high-power density ultra-capacitors for top-end consumer appliances.

1. Introduction

The properties of metals in nano-particulate form are currently receiving much attention. However, the inverse situation, that of a mesoporous metal skeleton containing nano-scale cavities, has scarcely been investigated. Nevertheless, such mesoporous materials have been known since at least 1927, when Raney patented his famous nickel hydrogenation catalyst produced by de-alloying Ni_xAl_y with $NaOH$[1]. 'Raney copper' and 'Raney platinum' may be produced by an analogous process[2]. Generally the resulting nanoscale sponge is pyrophoric and must be handled with care. However, mesoporous gold is a rare exception to this rule, and it has recently been shown that non-combustible, mesoporous sponges may be produced by de-alloying an $AuAl_2$ intermetallic precursor[3,4]. The result is a skeletal gold network, characterized by a continuous network of channels of nanoscale dimensions. In this paper we will describe the properties and possible uses of this mesoporous gold sponge.

2. Anomalous properties of mesoporous gold

Bulk gold is yellow and melts at 1064°C. However, these 'facts' do not apply to the sponge. For example, the melting point of nano-structured gold protrusions or particles of 5 nm radius is reduced by up to 400°C[5]. In combination with the well-known high mobility of Au surface atoms (due to their reluctance to bond with atmospheric gases), the net effect is that mesoporous gold sponges are relatively unstable with regard to sintering. When freshly prepared mesoporous gold is black, the state is one of the best absorbers known of infra red radiation[6].

However, unless the mesoporous state is stabilized by some coating substance, such as sodium citrate, aluminum oxide or copper oxide, the black gold has a tendency to sinter to the denser yellow form.

The size distribution of the channels in a sample with a BET surface area of $20 \text{ m}^2/\text{g}$ is shown in Figure 1. It is evident that this material consists of a large number of nominally 13 nm diameter cavities embedded in a matrix of gold.

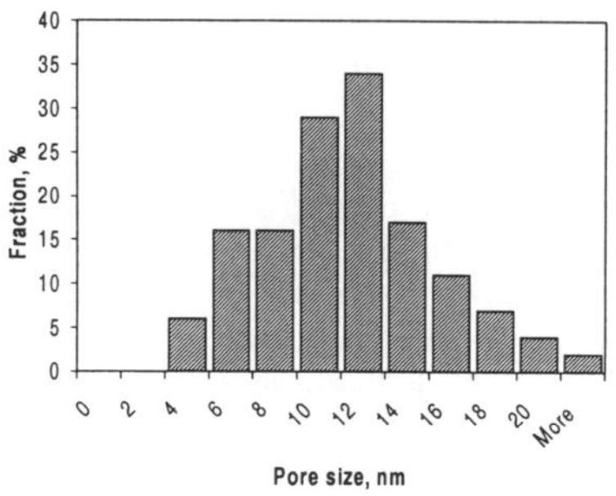

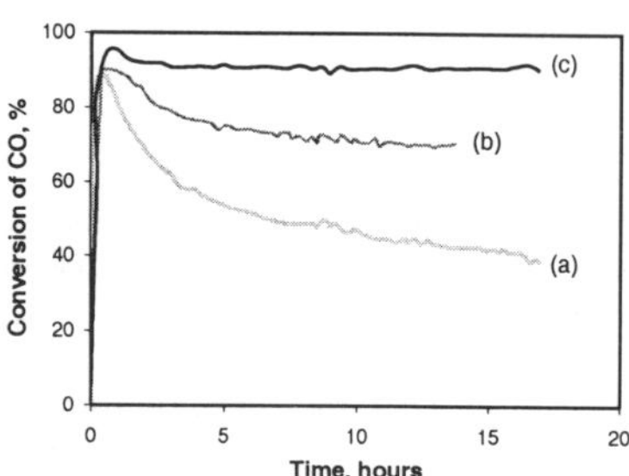

Fig. 1. Channel sizes in meso-porous gold surface prepared by de-alloying of $AuAl_2$ in 1M NaOH. Sample has been dried out for SEM examination but has been ostensibly stabilized by about 5 wt% Al_2O_3.

Fig. 2. Oxidation of 1% carbon monoxide in air at 60°C. The space velocity is 9 l.g^{-1}.h^{-1}; (a) commercial hopcalite, (b) hopcalite on which gold nanoparticles have been deposited, (c) mechanically alloyed composite of hopcalite and mesoporous gold.

3. Catalysis

It was only discovered in the 1980s that gold nano-particles can be active heterogeneous catalysts[7,8]. On the face of it, the nobility of metallic gold makes it an unlikely candidate for heterogeneous catalysis. However, gold particles of less than 10 nm diameter have now been shown to be active catalysts for diverse chemical reactions[9], of which the oxidation of carbon monoxide at low temperatures has been especially well studied[10]. The mechanism is still controversial[5] although the majority of opinion opts for the effect requiring a discrete gold nano-particle, an oxide support, and a special nanoscale electronic state. However, in contradiction to this we show that mesoporous gold networks are active catalysts too, especially if combined into a nano-composite with oxides such as Al_2O_3, Fe_2O_3, TiO_2 and Co_2O_3.

Pure mesoporous gold is not particularly active for CO oxidation, but the prior alloying of Fe into the $AuAl_2$ precursor produces a very active catalyst[4]. In this case the alloyed Fe^0 is converted to Fe_2O_3 during the catalysis of CO oxidation, producing an *in situ* oxide support. Alternatively, the $AuAl_2$ can be mechanically alloyed with a catalytic support material. Hopcalite ($Mn_xCu_yO_z$), for example, has long been known as a capable of oxidizing CO at room temperature. However, it deactivates within a few hours. Composite gold-hopcalite catalysts prepared by mechanical alloying followed by leaching retain their activity with regard to CO oxidation for very long times (Figure 2).

Another application for mesoporous gold is in the conversion of NO_x to nitrogen. There is no place here to discuss the intricacies of this process, save to note that it is a tough task to reduce NO_x to nitrogen in a gas stream that is super-stoichiometric in oxygen. In contrast to the situation with CO oxidation, the selective catalytic reduction of NO_x with propene occurs most readily on the unalloyed sponge[11] (Figure 3). The oxidation of NO to NO_2 can also be catalyzed, although in this case a gold-Co_2O_3 composite has proven to be the more active[11] (Figure 3).

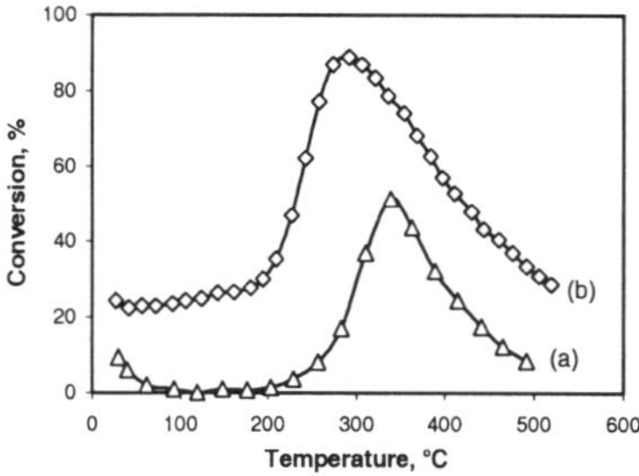

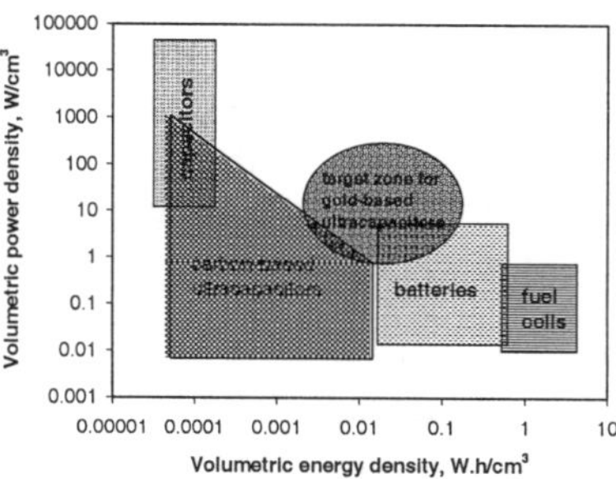

Fig. 3. (a) Selective catalytic reduction of NO to N2 over mesoporous Au with propene, (b) catalytic oxidation of NO to NO2 over mesoporous Au/CoxOy composite. The space velocity in both instances is 60l.g-1.h-1.

Fig. 4. Volumetric 'Ragone plot' showing the energy/power performance of various storage devices, and the anticipated niche for gold-based ultracapacitors. Based on use of an organic electrolyte with 2.5 V cell voltage.

4. Capacitance

The surface of a charged conductor placed into an liquid containing ions becomes coated with a double-layer of charged ions. If the surface area is very large, as for example for activated carbon at ~ 1000 m^2/g, then the capacitance can be very large. This effect has recently been exploited by some companies to produce 'ultra-capacitors'. Ultra-capacitors have considerably better power buffering capacities than batteries or ordinary capacitors but are not as good as batteries for storing energy. After accounting for energy lost to internal heating, the peak power P that can be drawn from an ultra-capacitor while discharging it from voltage V_0 to $\frac{1}{2}V_0$ is given by [12]:

$$P = \tfrac{9}{16}\left(1 - E_f\right)\frac{V_0^2}{R} \qquad (1)$$

where E_f is the discharge efficiency, and R is the internal resistance. The effect of this is that the power that can be supplied at a given efficiency is inversely proportional to the internal resistance. Since devices based on mesoporous gold should have very much lower resistances than carbon-based ultra-capacitors, they should deliver much higher volumetric power density. The maximum energy storage possible in a gold-based ultra-capacitor is not yet known. However, a electrolytic capacitance of 20 to 40 $\mu F/cm^2$ in combination with a surface area of 20 m^2/g, a density of ~7 g/cm^3 and a rapid discharge of 3 to 5 seconds implies that possible power densities of between 1 and 9 W/cm^3 are feasible, in comparison to the 3 W/cm^3 currently available from large carbon based ultra-caps[13]. The possible position of gold-based ultra-capacitors in terms of volumetric power density and energy storage is shown in Figure 4.

5. Conclusions

Mesoporous gold can be produced with a surface area of as high as 20 m^2/g but is susceptible to sintering unless stabilized by an admixture of oxide or other substance. It is an active catalyst for the selective catalytic reduction of NO_x, and, when combined with Fe, an active catalyst for the oxidation of CO. Its high surface area, electrical conductivity and corrosion resistance also suggest its use in ultra-capacitor technology, with volumetric power densities of between 1 and 9 W/cm^3 appearing to be achievable.

Acknowledgement
This work was supported by Project AuTEK and the World Gold Council.

6. References

1. M. Raney, *US Patent 1628190*, 1927.
2. J.B. Friedrich, D.J. Young, and M.S. Wainwright, *J. Electrochem. Soc.*, **128**, 1845 (1981).
3. E. Van der Lingen, M.B. Cortie and L. Glaner, *South African Patent 2001/5816 (2001)* & PCT patents pending.
4. E. van der Lingen, M.B. Cortie, H. Schwarzer, S.J. Roberts, G. Pattrick, *Gold 2003*, 28th September-1st October 2003,Vancouver, Paper 988.
5. M.B. Cortie and E. Van der Lingen, *Materials Forum*, **26**, 1 (2002).
6. D.J. Advena, V.T. Bly, J.T. Cox, *Applied Optics*, **32(7)**, 1136 (1993).
7. G.J. Hutchings, *J. Catalysis*, **96**, 292 (1985).
8. M. Haruta, T. Kobayashi, H. Sano, and M. Yamada, *Chemistry Letters*, p.405 (1987).
9. D. Thompson, *Gold Bulletin*, **31(4)**, 111 (1998).
10. D. Thompson, *Gold Bulletin*, **32(1)**, 12 (1999).
11. G. Pattrick, E. van der Lingen, H. Schwarzer and S.J. Roberts, *Gold 2003*, 28th September-1st October 2003,Vancouver, Paper 993.
12. A. Burke, *J. of Power Sources*, **91**, 37 (2000).
13. M. Cortie and E. van der Lingen, *Gold 2003*, 28th September-1st October 2003,Vancouver, Paper 1064.

FRONTIERS OF PHASE FORMATION AND CONTROL IN NANOPARTICLES BY *IN-SITU* TEM[*]

H. YASUDA

*Department of Mechanical Engineering, Kobe University, Rokkodai, Nada,
Kobe 657-8501, Japan*

H. MORI AND J. G. LEE

*Research Center for Ultra-High Voltage Electron Microscopy, Osaka University,
Suita, Osaka 565-0871, Japan*

Phase control in nanomaterials is one of the key technologies for nanofabrications. Recently, we have found by *in-situ* transmission electron microscopy that when electronic excitations by 75 keV electrons were carried out in GaSb particles, the original compound crystalline phase changes to the two-phase consisting of an antimony core and a gallium shell or an amorphous phase depending on the temperature of the specimens and the particle size. Using the synergistic responses of phase changes by electronic excitation, we can control the phase formation in nanomaterials.

1. Introduction

In-situ transmission electron microscopy is one of useful methods to investigate atom displacements induced by both self-organization and electron irradiation in nanomaterials such as in nanoparticles. Using this technique, we have observed directly that atom displacements driven by the free energy difference between the initial atomic structure and the final atomic structure could be induced in nanoparticles of binary systems in which the heat of mixing is negative with ease(1-3). It has been considered as the mechanism that the structure is successively self-organized to minimize the chemical free energy in each chemical composition, because the activation barriers for the atom displacements and reactions become quite low by characteristic lattice softening in nanoparticles (4,5).

On the other hand, by electron irradiation techniques in an electron microscope we can control atom displacements which may be introduced as a response to alteration of atomic electronic states by ionizing radiation (6). Interesting responses of phase changes may be expected to appear by irradiation effects in nanomaterials, since the electronic excitations tend to localize remarkably in nanomaterials which are the isolated system. Recently, we have found synergistic (nonlinear) responses of phase changes driven by alteration of

---

[*] This work is supported by the Ministry of Education, Culture, Sports, Science and technology under Grant-in-Aid for Scientific Research.

electronic states induced by electronic excitations localized in III-V compound nanoparticles such as GaSb particles.

2. Synergistic responses of phase changes induced by electronic excitations

Examples of phase changes in GaSb particles induced by electronic excitation are shown in Fig. 1, as functions of temperature of the specimens during excitation and the particle size.

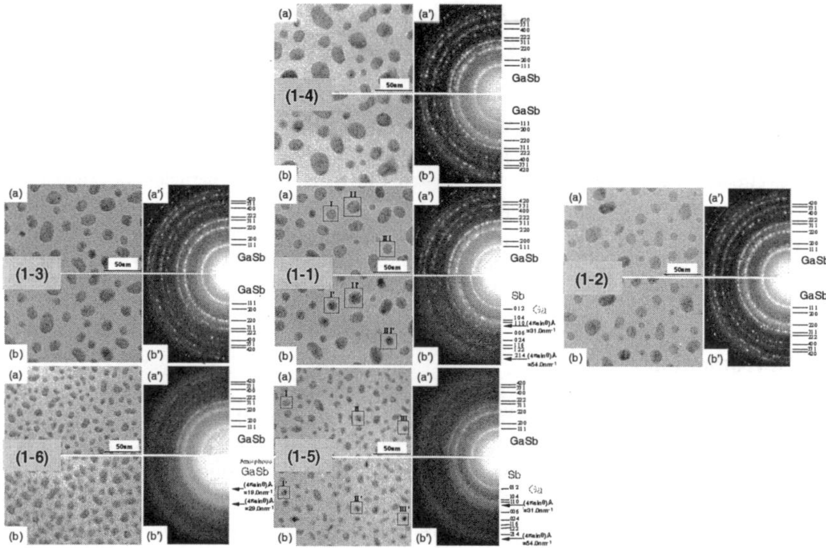

Figure 1. Phase changes in GaSb particles induced by electronic excitation with 75 keV electrons under various conditions. (a) and (a') indicate a BFI and the corresponding SAED before excitation, respectively. (b) and (b') indicate a BFI and the corresponding SAED after excitation for 240 s, respectively. (1-1) Approximately 20 nm-sized particles at 423 K. (1-2) Approximately 20 nm-sized particles at 443 K. (1-3) Approximately 20 nm-sized particles at 293 K. (1-4) Approximately 30 nm-sized particles at 423 K. (1-5) Approximately 10 nm-sized particles at 423 K. (1-6) Approximately 10 nm-sized particles at 293 K.

An example of the phase change by electronic excitation in GaSb nanoparticles kept at 423 K is shown in Fig. 1-1. Figures 1-1(a) and (a') show a BFI of particles with the diameter of approximately 20 nm before excitation and the corresponding SAED, respectively. As indexed in the figure (1-1(a')), the Debye-Scherrer rings can be consistently indexed as those of GaSb (which has the zincblende structure with lattice constant of a_0=0.61 nm). The same area after excitation for 240 s is shown in Fig. 1-1(b). In the interior of particles after the excitation, there appears a structure consisting of a core with dark contrast and a shell with bright contrast, as seen from a comparison of the parts framed by I, II and III in (a) with those framed by I', II' and III' in (b), respectively. The corresponding SAED taken after the excitation is shown in Fig. 1-1(b'). In the SAED, Debye-Scherrer rings are recognized, superimposed on halo rings. The Debye-Scherrer rings can be indexed consistently as those of crystalline antimony which has the hexagonal structure with lattice constants of a_0 =0.43 nm and c_0 =1.13 nm. The values of the scattering vector ($K=(4\pi\sin\theta)/\lambda$) for the halo rings are approximately 31.0 nm^{-1} and 54.0 nm^{-1} which are corresponding to those from liquid gallium. This fact indicates that a two-phase mixture of crystalline antimony and liquid gallium is formed in the particles after the excitation. It was confirmed by dark-field electron microscopy that nanoparticles after the electronic excitation are the two-phase structure consisting of a crystalline antimony core and a liquid gallium shell.

The temperature dependence of the behaviours by electronic excitation in approximately 20 nm-sized GaSb particles kept at 443 K and 293 K has been examined as shown in Fig. 1-2 and 1-3, respectively. It is noted here that any phase changes are absent during the electronic excitation also in the particles kept at 443 K and 293 K.

The particle-size dependence of the behaviours by electronic excitation in nanoparticles kept at 423 K has been examined using both larger (i.e., 30nm-sized) and smaller (i.e., 10nm-sized) particles than those in Fig. 1-1. As shown in Fig. 1-4, a phase separation is not induced with increasing size to approximately 30nm. However, it is evident that the phase separation by the electronic excitation is induced also in approximately 10nm-sized particles kept at 423 K as shown in Fig. 1-5.

In order to see the temperature dependence of the phase change in approximately 10nm-sized particles, an electronic excitation experiment in GaSb nanoparticles kept at 293 K was carried out. An example of the phase change is shown in Fig. 1-6. In the SAED taken from the region irradiated for 240 s (Fig. 1-6(b')), only haloes are recognized. The values of the scattering vector for the halo rings are approximately 19.0 nm^{-1} and 29.0 nm^{-1} which are corresponding to those from amorphous GaSb. This result indicates that amorphization has been induced by the excitation in the 10 nm-sized GaSb particles kept at 293 K.

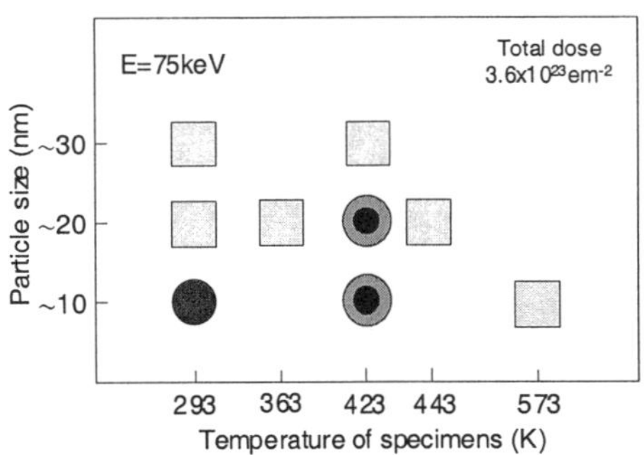

Figure 2. Summary of the results on phase changes induced by electronic excitation as functions of temperature of the specimens during excitation and the particle size. The marks of a blue and yellow duplicate circle (◉), a red circle (●) and a light blue square (☐) indicate the two-phase separation consisting of a crystalline antimony core and a liquid gallium shell, the formation of the amorphous phase and an original crystalline phase with the zincblende structure, respectively

Figure 2 summarizes the results on phase changes induced by electronic excitation as functions of temperature of the specimens during excitation and the particle size. Total dose of electrons are fixed at 3.6×10^{23} e m^{-2}. In this figure, marks of a blue and yellow duplicate circle, a red circle and a light blue square indicate the two-phase separation consisting of a crystalline antimony core and a liquid gallium shell, the formation of the amorphous phase and an original crystalline phase with the zincblende structure, respectively. Through the present experiments, it was evident that when GaSb nanoparticles have been electrically excited by 75 keV electrons, synergistic responses of phase changes appear depending on the temperature of the specimens and the particle size.

3. Driving force of the phase changes under electronic excitations

It is well-known as a mechanism that the electronic excitation-induced atom displacements occur due to breaking of the bond by local electronic transitions i.e., the formations of electron-hole (e-h) pairs or pairs of holes (h-h) (7, 8). A similar mechanism could play an important role also in phase changes induced by electronic excitations in GaSb nanoparticles. It is considered that atom displacements could be remarkably enhanced, since multiple e-h or h-h pairs produced by the electronic excitations tend to localize remarkably in nanoparticles. The atom displacements may lead to induce the resulting phase separation or amorphization. It is suggested that the phase changes may arise from synergistic effects of bond instability by electronic excitation, localization of the excited states in nanoparticles, enhanced atomic mobility causing the lattice softening characteristic of nanoparticles or thermal equilibrium and kinetics in the reactions. In conclusion, when GaSb nanoparticles have been electrically excited, nonlinear responses of phase changes such as phase separation and amorphization appear depending on the particle size and specimen temperature. The electronic excitations using an electron beam could be an important technique to control phase changes reversibly in nanomaterials.

References

1. H. Yasuda, and H. Mori, *Phys. Rev. Lett.* **69**, 3743 (1992).
2. H. Yasuda and H. Mori, *Z. Phys.* **D31**, 131 (1994).
3. H. Yasuda and H. Mori, *Thin Solid Films* **298**, 143 (1997).
4. H. Yasuda, H. Mori and K. Furuya, *Phil. Mag. Lett.* **80**, 181 (2000).
5. H.Yasuda, K. Mitsuishi and H. Mori, *Phys. Rev.* **B64**, 094101-1 (2001).
6. A. Schmid, P. Braunlich and P. K. Rol, *Phys. Rev. Lett.* **35**, 1382 (1975).
7. J.Kanasaki *et al.*, Phys. Rev. Lett., 70, 2495 (1993).
8. J. Singh *et al.*, Phys. Rev. B, 50, 11370 (1994).

TEMPLATE DIRECTED SYNTHESIS OF NANOSIZED BONE-LIKE APATITE

Adriyan S. Milev [a], G. S. Kamali Kannangara [a],
Besim Ben-Nissan [b] and Michael A. Wilson [a]

[a] *College of Science, Technology and Environment, University of Western Sydney,*
Locked Bag 1797, Penrith South DC 1797, Australia.
[b] *Department of Chemistry, Materials and Forensic Science, University of Technology, Sydney, PO Box 123, Sydney 2007, Australia.*

ABSTRACT

In order to maximize the bioactivity of prosthetic materials, synthesis of nanosized hydroxyapatite is required. In addition, it is highly desirable the synthetic hydroxyapatite to have similar chemical substitutions and morphology of biological apatites. A novel method has been developed to produce single phase, nano sized, plate-like, mixed A-B type carbonate containing apatite (CAp) similar to bone apatite for effective bone tissue integration. The methodology emulates biomineralization, where topotactic transition from octacalcium phosphate (OCP) to hydroxyapatite (HAp), which is believed to occur *in vivo*. The process involves formation of thin (~1.4 nm) layered calcium phosphonate salts by a self-assembly process. The thermal decomposition of these layered salts leads to formation of plate-like carbonated apatite. The overall carbonate content varies from 6.4 to 4 wt%, within the temperature range of 500 – 700 °C. This carbonate content corresponds well with the amount found in mammalian hard tissues.

KEYWORDS: Hydroxyapatite, phosphonate, carbonate, morphology.

INTRODUCTION

Bone material consists of poorly crystallized hydroxyapatite $[Ca_{10}(PO_4)_6(OH)_2]$ containing carbonate ions.[1] Biogenic apatite is believed to be formed from octacalcium phosphate (OCP) *via* stoichiometric protons loss and substitution of calcium ions for protons (Eq 1). [1-4]

$$Ca_8H_2(PO_4)_6 \cdot 5H_2O + 2Ca^{2+} \rightleftharpoons Ca_{10}(PO_4)_6(OH)_2 + 4H^+ + 3H_2O \quad (Eq\ 1)$$

The process is topotactic.[2, 3] That is, transformation of OCP to apatite with no change of the morphology. During transformation carbonate ions may be incorporated into the apatite crystallites.[4] The incorporation may be of two types. When a carbonate anion occupies lattice positions in the OH^- sublattice it is named A-type carbonate apatite. Carbonate can also be incorporated *via* phosphate displacement. This is called B-type substitution. Bone apatite contains CO_3^{2-} in both types of locations in ratio A to B of 0.7 to 0.9. [5, 6] The crystallites of bone apatites are plate shaped nano sized particles with lengths and widths of 50 x 25 nm and thickness of 2 – 5 nm.[7] Hence, synthetic strategies to produce bone apatite confront two challenges; to incorporate similar proportions of carbonate in both A and B positions and to mimic morphology.

In our previous work we investigated the chemical reactions in solution at 70 °C aimed at hydroxyapatite synthesis.[8] An intermediate product was shown to contain phosphonate chelates. Interestingly after subsequent drying at 130 °C this solution formed a white powder, which has lamellar morphology like biogenic hydroxyapatite. The thermal decomposition of the powder at temperatures higher than 500 °C led to synthesis of single-phase hydroxyapatite in which the plate-like morphology of the phosphonate precursor was preserved.[9] Thus, the plate-like calcium phosphonates act as a template for apatite growth in retaining the precursor morphology.

In this paper we discuss the incorporation of carbonate into the product, which originates from the organics of the phosphonate intermediates and show that this process mimics biological systems.

EXPERIMENTAL

Materials, Sample preparation, Characterization. Calcium diethoxide $Ca(OC_2H_5)$ > 99% purity, diethyl hydrogen phosphonate $H(O)P(OC_2H_5)_2$ > 98% purity were mixed in stoichiometric ratio of 1.67 in a mixed solvent containing acetic acid > 99.7% purity and ethylene glycol > 99.5% purity in 1: 1 molar ratio. The starting solution was heated at 70 °C for 48 h in a closed vial. The reaction products were solidified by evaporation at 130 °C in a Petri dish. Hydroxyapatite samples were then obtained after firing of the precursor powders at temperatures ranging from 500 to 900 °C for 2 h. Infrared spectroscopy (IR) and X-ray diffraction data are used to characterize the type and the amount of the carbonate in hydroxyapatite lattice and the average sizes of the hydroxyapatite crystallites. Scanning Electron Microscopy (SEM) was used to characterize the particles morphology.

RESULTS AND DISCUSSION

Scanning Electron Microscopy

The thermal decomposition of the precursor powder at temperatures higher than 500 °C led to synthesis of single-phase hydroxyapatite particles in which the plate-like morphology of the phosphonate precursor was preserved. Thus, the plate-like calcium phosphonates act as a template for apatite growth in retaining the precursor morphology (Figure 1).

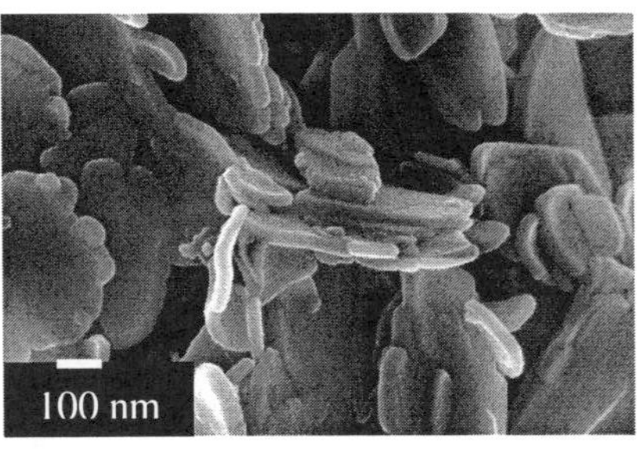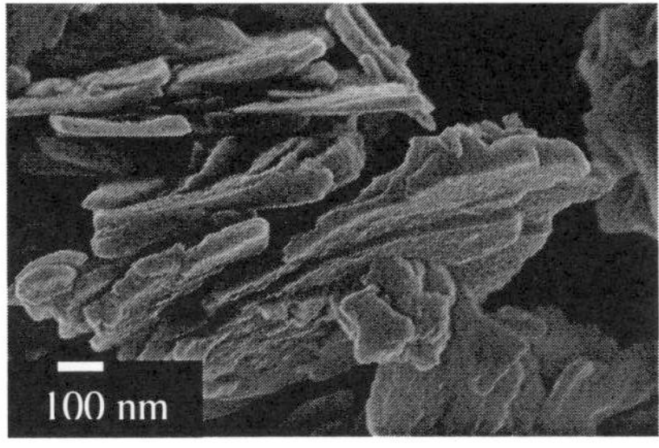

A   B

Figure 1. SEM of hydroxyapatite precursor particles obtained at 130 °C (A) and hydroxyapatite particles obtained at 500 °C (B).

X-ray diffraction

The average crystallite sizes may be approximated by well-known Debye-Scherrer formula based on the assumption that the lattice strain is negligibly small.[10] The average sizes of the hydroxyapatite crystallites were found to range from about 30 nm (500 °C) to about 50 nm (900 °C). These values fit well with the reported data for biogenic apatites. It should be noted that the particle sizes determined microscopically are higher than that obtained by X-ray diffraction. It may be due to the fact that X-ray diffraction is sensitive to the crystallite size, while SEM provides particle size information.

The inclusion of CO_3^{2-} is known to impose changes in the lattice parameters.[11] This variation in the lattice is dependent on the type of substitution. It is reported that Type A (CO_3^{2-} for OH^-) causes expansion of the a- and the contraction of the c-axis dimensions[12], while type B carbonated apatite (CO_3^{2-} for PO_4^{3-}) causes the opposite variation. i.e. contraction of the a- and expansion of the c- axis dimensions.[13]

The phosphonate-derived apatite shows significant changes of both a- and c-axis as function of temperature (Figure 2 A). At lower temperatures (500 – 600 °C) both a- and c- lattice constants have higher values than that of stoichiometric hydroxyapatite (a-axis = 0.94180 nm and c-axis = 0.68840 nm, JCPDS standard, File Card No. 9-432) due to the formation of type A and B carbonated apatites. With the increase of the temperature a- and c- lattice constant values gradually decrease reaching a minimum at 800 °C, which is attributed to release of some of the carbonate. The increase at 900 °C is most likely caused by the breakdown of B- type carbonate, release of CO_2 and formation of CaO phase. It can be envisaged that the crystal lattice expands and that the a- axis increases due to liberation of carbon dioxide and rearrangement of the structure to form mostly type B apatite.

90

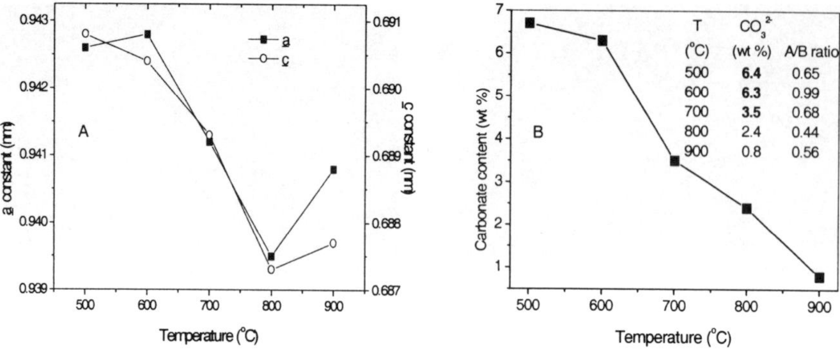

Figure 2. A- The apatite lattice constants $\underline{a}$- and $\underline{c}$-axis as a function of temperature. Peaks corresponding to 002 and 310 planes are used to calculate the lattice parameters. B- Estimation of the carbonate contents and type carbonate ratios (A to B ratios) as function of temperature.

We note that due to the simultaneous presence of carbonate ions in two structural positions in apatite the variations to the lattice parameters along $\underline{a}$- and $\underline{c}$-axis would result in a mutual cancellation[14] and therefore have not been used for quantitative determination of carbonate substitutions. However, the analysis of the data does suggest there is greater populations of B-type carbonate substitution than A-type carbonate.

Infrared spectroscopy

The carbonate content of apatites has been estimated using IR method exploiting CO_3^{2-}/PO_4^{3-} extinction coefficients of the ν_3 carbonate band at ~ 1412 cm^{-1} using the ν_4 phosphate band near 575 cm^{-1}. It was shown that this normalization is linearly related to the carbonate contents in the range $1 - 12$ wt%.[15] The calculated carbonate contents for the phosphonate-derived apatite are given in Figure 2 B. Significant decrease in the carbonate content occurs between 600 and 800 °C, which further support the XRD data. The apatite can be classified as carbonate apatite with carbonate substitution in both OH$^-$ (A-type) and PO_4^{3-} (B-type) structural positions with carbonate content in the temperature interval $500 - 900$ °C ranging from $6.4 - 0.8$ wt %.

The ν_2 carbonate bands appear in $880 - 870$ cm^{-1} region may be assigned to two different locations of the ion in the mineral: in monovalent anionic sites (A-type 880 cm^{-1}), and in trivalent anionic sites (B-type 873 cm^{-1}). A comparison of the intensities at ν_2 domain of A- and B-type carbonates, shows that there is a greater degree of carbonate substitution at phosphate (B) site than on hydroxyl (A) site for all temperatures except for the sample treated at 600 °C where such a preference is not observed. It is interesting to note that the magnitude of $\underline{a}$-axis, which is considered to be sensitive to A-type carbonate, is at its maximum at 600 °C (Figure 2 A).

It is remarkable that the carbonate content and A- and B- type carbonate ratios at temperatures up to 700 °C agree well with the corresponding values found in mammalian hard tissues. The carbonate content, the types of substitution and the morphology of the apatite phase all suggest that the synthetic apatite obtained from this procedure mimics biological hard tissues. The CO_3^{2-} concentration is least in human enamel apatites (~3.5 wt%) and most in bone (~7.4 wt%) [16] and the values described here fall within range. Likewise, the substitution patterns of B- type *vs* A-type substitution observed here are common in mammalian bone mineral.

CONCLUSIONS

1. Bone-like hydroxyapatite has been produced through lamellar intermediate phase.

2. The hydroxyapatite obtained *via* the phosphonate chelate intermediates contains carbonate. The percentage of carbonate in the product depends on the thermal treatment temperature range 500 – 900 °C.

3. Carbonate substitutes for both PO_4^{3-} an OH^- in the apatite lattice. A comparison of IR absorption intensities shows that there is a greater degree of carbonate substitution at phosphate sites than on hydroxyl sites for all temperatures except for the sample treated at 600 °C, which does not show a preference towards sites. X-ray diffraction data also supports this conclusion.

4. The degree of substitution and carbonate content in the synthetic apatite is similar to mammalian hard tissues and demonstrate that the material is worth exploiting in animal trials for bone replacement or implant uses.

REFERENCES

1. Posner, A.; Betts, F.; Blumental, N. *Bone mineral composition and structure.* in Skeletal research: Academic Press, 1979: 167.
2. Brown W. *Nature* 1962; **196**: 1048.
3. Brown W., Schroeder L., Ferris J. *J Phys Chem* 1979; **83**: 1385.
4. Boskey A., Posner A. *Structure and formation of bone mineral*, in Natural and Living Biomaterials; Boca Raton: CRC Press, 1984: 27.
5. Emerson W., Fisher, E. *Arch Ora. Biol* 1962; **7**: 671.
6. Biltz R., Pellegrino E. *Clin OrthopRelat Res* 1977; **129**: 279.
7. Robinson A. *J Bone Joint Surg* 1952; **34A**: 389.
8. Milev A., Kannangara G. S. K., Ben-Nissan B., Wilson M. *J Phys Chem* **Submitted.**
9. Milev A., Kannangara G. S. K., Ben-Nissan B. *Materials Letters* 2003; **57**: 1960.

10. Klug H., Alexander I. *X-ray Diffraction Procedures for Polycrystalline and Amorphous Materials.* New York: Wiley & Sons, 1974.
11. LeGeros R. *Nature* 1965; **206**: 403.
12. Binder G., Troll G. *Contrib Mineral Petr* 1989; **101**: 394.
13. LeGeros R., LeGeros J., Trautz O., Shirra W. *Adv X-ray Anal* 1971; **14**: 57.
14. Xu G., Aksay I., Groves J. *J Am Chem Soc* 2001; **123**: 2196.
15. Featherstne J., Pearson S., LeGeros R. *Caries Res* 1984; **18:** 63.
16. Nainwright S., Briggs W., Currey J., Gosline J. *Mechanical Design in Organisms.* Princeton: Princeton University Press, 1976.

USING CARBON NANOTUBES AND DNA TO TRANSFER ELECTRONS OVER LONG DISTANCES: APPLICATIONS TO SENSING AND BIOELECTRONICS [*]

J. JUSTIN GOODING, JINGQUAN LIU, ELICIA (LEE-SEH) WONG, WIBOWO RAHMAD, ALISON CHOU, MICHAEL N. PADDON-ROW, D. BRYNN HIBBERT

School of Chemical Sciences, The University of New South Wales
Sydney, NSW, 2052, Australia

DUSAN LOSIC, JOE G. SHAPTER

School of Chemistry, Physics and Earth Sciences, Flinders University
Adelaide, 5001, Australia

Two methods of exploiting the transport of electrons over long distances for sensing applications are described. In the first approach carbon nanotubes are aligned normal to an electrode surface via self-assembly and subsequently enzymes are attached to the ends of the aligned tubes. As each tube can act as a molecular wire, efficient communication with the redox active centre of the enzymes is achieved despite the enzyme being located a 100 nm or more from the underlying electrode surface. In the second method, the ability of a double strand of DNA to transfer electrons over long distances is exploited to detect DNA hybridization with the ability to differentiate between a perfectly complementary sequence and a target sequence containing a single base pair mismatch.

1. Introduction

Often bionanotechnology and bioelectronics involves the integration of biological molecules with electronic elements and is of intense interest for biosensing, creating electronic readouts of biomolecular function, the assembly of nanocircuit elements and the conversion of biocatalytic processes into electrical power [1]. A crucial step in all these applications is the transfer of electrons to and from the biological molecule. In many instances however the redox centre of the protein of interest is embedded deep within the tertiary structure of the polypeptides [2]. Effective communication to such proteins could be achieved by building conduits for electron movement from the

---

[*] This work is supported by the Australian Research Council and The University of New South Wales

electrode up into the protein. Such conduits could be molecular wires (electron transport) or molecules which provide efficient tunneling of electrons over long distances (electron transfer).

With regards to molecular wires, carbon nanotubes are ideal [3] as they have good electron transport properties to redox enzymes [4, 5]. Furthermore, because of their small size, carbon nanotubes could potentially be brought up inside proteins, close to the active site, thus allowing efficient electrical communication with proteins in which it is difficult to achieve direct electron transfer [6]. In contrast, double strands of DNA have been shown to transfer electrons over long distances [7]. The mechanism of charge transport is believed to be either via super-exchange or charge hopping between localized sequences in the double helix [8]. The ability of double strands of DNA to transfer electrons over long distances and not single strands means that charge transfer through DNA is ideal for detecting DNA hybridization with sensitivity to base pair mismatches [9]. In this paper, both the application of single walled carbon nanotubes (SWNT) for communicating with enzymes with a view to fabricating a enzyme sensor and the change in the ability of DNA to transfer electrons over long distances upon hybridization as the basis of a DNA biosensor are demonstrated.

2. Results and Discussion

2.1. *Carbon Nanotubes for Communicating with Enzymes*

The long term goal of this work is to use carbon nanotubes to allow direct electrical communication with enzymes (which essentially means turning the redox enzyme over directly at an electrode) which would not normally allow communication with their redox centre due to the depth within the glycoprotein that the redox active centre is embedded. An example of such an enzyme is glucose oxidase where the flavin adenine dinucleotide (FAD) active centre is embedded 13 Å within the glycoprotein. Before investigating such enzyme systems it was first important to demonstrate that nanotube electrode arrays for communicating with enzymes could be fabricated using simpler enzymes where the redox active centres were close to the enzyme surface such that they could normally be oxidised and reduced at an electrode. For this purpose the enzyme microperoxidase MP-11 was used which a small redox protein (1.9 kDa) is obtained by the proteolytic digestion of horse heart cytochrome c [10].

The method of fabricating aligned nanotube electrodes with enzymes attached to the ends is shown in Fig 1. The uncut SWNTs, which resembled tangled hair in the TEM (see Fig. 1a) are oxidatively shortened to give a log normal distribution of tube lengths (Fig. 1b) and then covalently attached to a gold surface modified with a self-assembled monolayer of cysteamine via carbodiimide coupling (Fig. 1c). Finally the MP-11 is covalently attached to the other ends of the aligned SWNTs using the same carbodiimide coupling procedure (Fig. 1d) [11].

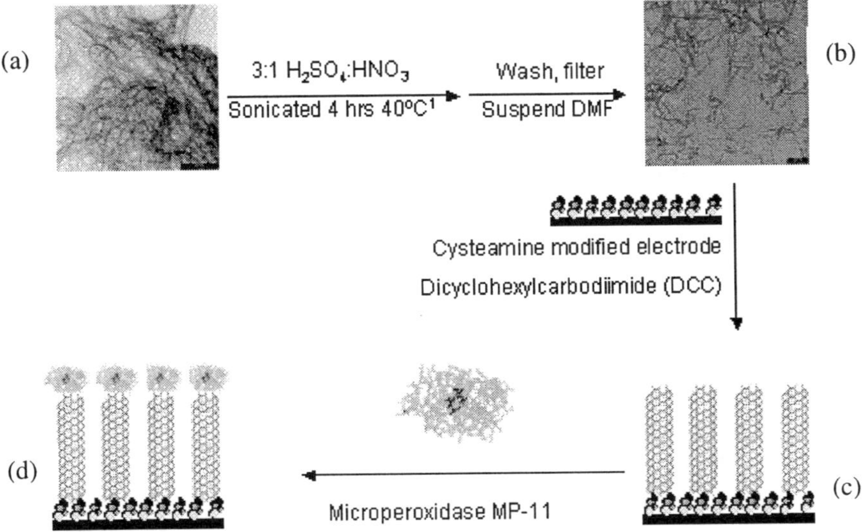

Figure 1. The procedure for the fabrication of enzyme modified aligned carbon nanotube arrays.

Prior to electrochemical testing of these interfaces it was important to verify that the electrode constructed with aligned nanotubes represented in Fig.1 is realized in reality. For this purpose tapping mode atomic force microscopy was employed. AFM images of these interfaces after the assembly of the tubes are shown in Fig.2 below. Figure 2 unambiguously shows that the assembly process produces SWNT aligned normal to the self-assembled monolayer modified electrode surface with some tubes aligned parallel to the surface.

Attaching MP-11 to the aligned SWNT modified gold electrodes and subsequent electrochemical interrogation showed the characteristic peaks for the heme redox active center of MP-11 with formal electrode potential of – 420 mV versus Ag/AgCl. In the absence of MP-11 no such electrochemistry was observed, nor was it observed when the electrode was modified with a bed of SWNT such that few ends were exposed. The rate of electron transfer from the electrode, through the nanotubes to MP-11, was calculated to be 2.8 ± 0.9 s^{-1} for tubes cut for four hours which is similar to the rate calculated when the MP-11 is attached directly to the cysteamine SAM (6.1 ± 0.9 s^{-1}) [5]. The similarity in rate constant despite the fact that the average length of the nanotubes in over 100 nm, indicates the efficiency of the SWNTs as molecular wires and their applicability as individual electrodes which allows efficient communication with redox proteins.

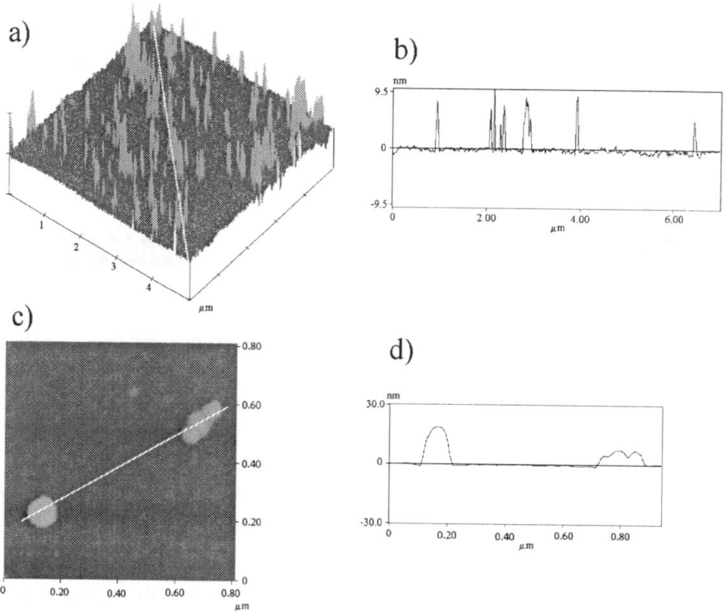

Figure 2. AFM images of shortened SWNTs aligned onto gold electrodes. a) tubes cut for 4 hours followed by activation with DCC and then incubated with a cysteamine modified gold electrode for four hours. b) cross section of part a) diagonally as shown, c) top view AFM image showing a bundle of aligned SWNTs and a bundle of tubes lying flat. The difference in heights between the standing tubes and those lying down is shown in part d).

2.2. *Electron Transfer through for Detecting DNA Hybridization*

The second example of the application of long range electron transfer to biosensing is in the development of DNA hybridization biosensors. DNA duplexes have been shown to allow electron transfer over distances of more than 40Å with both photoexcited [7] and ground state electrons [12]. From a sensing perspective the crucial observation is that perfect base pair stacking is required for rapid electron transfer [12] and therefore no electron transfer will be observed with ss-DNA and an diminution in the signal is expected with mismatched sequences [9, 12].

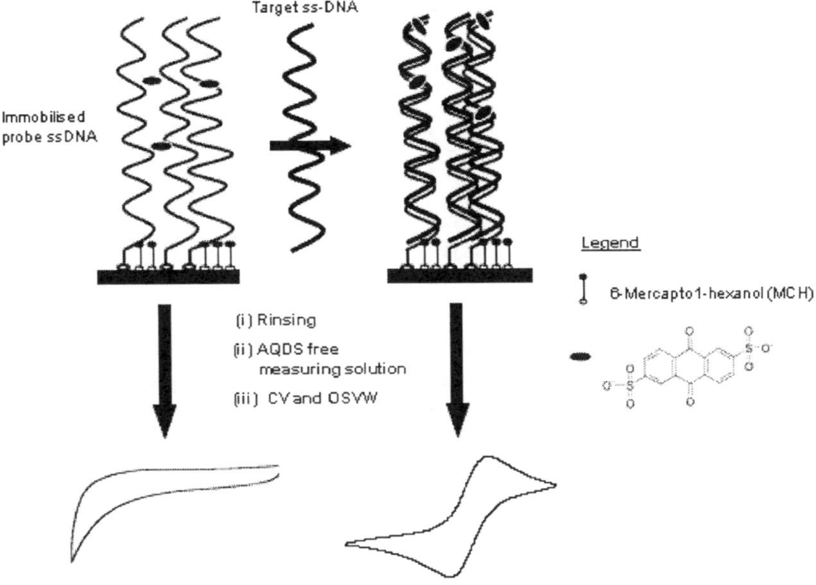

Figure 3. Schematic of the electrode construct for DNA hybridization detection using long range electron transfer through DNA duplexes and the measurement protocol.

Long range electron transfer based DNA hybridization biosensors were developed using the electrode construct shown in Fig 3. An important feature of the electrode construct was the diluent layer of MCH which served the dual function of firstly preventing the DNA bases adsorbing onto the gold electrode and orientating it out into solution and secondly preventing non-specific adsorption of the negatively charged electroactive intercalator AQDS. The ability of the diluent layer to perform this second function is shown in Fig 4(i) where no Faradaic current is observed for a ss-DNA modified electrode. Upon hybridization with the complementary sequence, the duplex is formed and electron transfer can proceed once the AQDS intercalates, Fig 4(ii). Exposure to noncomplementary sequences also gives no current, Fig 4(iii), whilst hybridization with either C-A mismatched sequences or the difficult to detect G-A mismatched sequences resulted in a least 70% diminution in current.

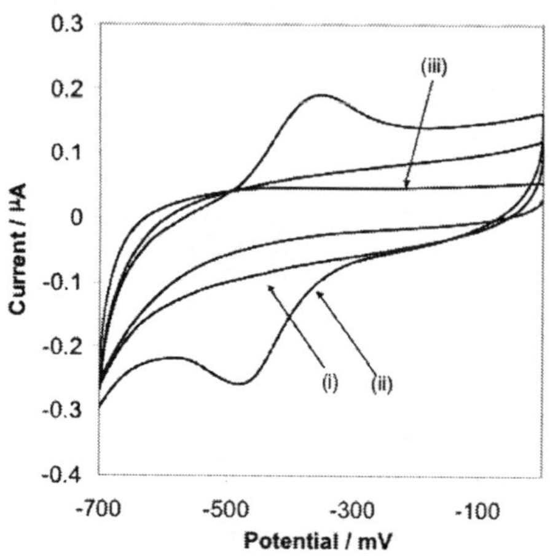

Figure 4. Cyclic voltammogram of (i) probe ss-DNA/MCH modified gold electrode; (ii) after exposure to complementary target ss-DNA; (iii) after exposure to non-complementary target ss-DNA.

3.   Conclusions

Advances in bionanotechnology afford unique opportunities for the development of new biosensing concepts. In this paper the bottom up fabrication of sensing interfaces which exploit the transfer of electrons over long distances has been demonstrated using carbon nanotubes interfaced with enzymes and with DNA for hybridization detection. Such strategies could eventually be applied to biosensors which can probe changes in the extracellular medium or on a subcellular level.

References

1. I. Willner, Science 298, 2407, (2002)
2. A. Heller, Acc. Chem. Res. 23, 128 (1990)

3. S. Niyogi, M.A. Hamon, H. Hu, B. Zhao, P. Bhowmik, R. Sen, M.E. Itkis and R. Haddon, Acc. Chem. Res. 35, 1105 (2002)
4. J.J. Davis, R.J. Coles and H.A.O. Hill, J. Electroanal. Chem. 440, 279 (1997)
5. J.J. Gooding, R. Wibowo, J.Q. Liu, W.R. Yang, D. Losic, S. Orbons, F.J. Mearns, J.G. Shapter and D.B. Hibbert, J. Am. Chem. Soc. 125, 9006, (2003)
6. A. Guiseppi-Elie, C.H. Lei and R.H. Baughman, Nanotechnology 13, 559 (2002)
7. R.E. Holmlin, P.J. Dandliker and J.K. Barton, Angew. Chem. Int. Ed. Engl. 36, 2714 (1997)
8. M.N. Paddon-Row, Aust. J. Chem. 56, 729, (2003)
9. E.L.S. Wong and J.J. Gooding, Anal. Chem. 75, 3845 (2003)
10. T. Lotzbeyer, W. Schuhmann, E. Katz, J. Falter and H.L. Schmidt, J. Electroanal. Chem. 377, 291 (1994).
11. J.J. Gooding and J.G. Shapter, Carbon Nanotube Systems to Communicate with Enzymes in Protein Nanotechnology , Humana Press, in press
12. E.M. Boon, D.M. Ceres, T.G. Drummond, M.G. Hill and J.K. Barton, Nature Biotech. 18, 1906 (2000)

AN ANALYTICAL SINGLE-ELECTRON-TRANSISTOR MODEL BASED ON FREE ENERGY DIFFERENCE ACROSS THE TUNNELING BARRIER [*]

YOU-LIN WU, and SHI-TIN LIN

Department of Electrical Engineering, National Chi-Nan University
301 University Rd., Puli, Nantou, Taiwan

In this paper, we propose an analytical single electron transistors (SET) model based on single-electron tunneling phenomenon, which is conventionally described by the orthodox theory. While most of the single electron transistor analytical models found in the literature focused on unilateral electron tunneling across the tunneling junctions, the proposed model uses free energy difference between energy levels on each side of the tunneling barriers and considers electron tunneling from either side of the barriers. The simulation results of the proposed model are quite close to those of the widely accepted Monte-Carlo SET simulator SIMON 2.0

1. Introduction

Since the operation of SETs involves only a few electrons, operation principles and device characteristics are quite different from those of conventional MOS transistors [1]. Modeling devices and simulating SETs are quite important in elucidating their characteristics. In general, simulation program based on Monte Carlo method [2-4] can give the most accurate results but long computation time is required. Despite the least computation time and its friendliness to those users who are not familiar with the device physics of SETs, the compact macro-model [5,6] is in lack of accuracy. On the other hand, analytical model [7,8] not only exhibits accurate results but also provides physical meanings behind the model and takes moderate computation time. In this paper, we proposed an analytical SETs model based on single-electron tunneling phenomenon, which is conventionally described by the orthodox theory. We used the free energy difference between energy levels on each side of the tunneling barriers and considered electron tunneling from either side of the barriers. A comparison with the SETs simulator SIMON 2.0 reveals that the proposed model can correctly describe the I-V characteristics of SETs.

[*] This work is supported by National Science Council of the Republic of China under Contract no. NSC 91- 2120-E-260-001

2. Analytical Model

The orthodox theory states the single-electron-tunneling rate as:

$$\Gamma = \frac{1}{e^2 \cdot R_T} \frac{-\Delta F}{1 - \exp(\Delta F / K_B T)} \tag{1}$$

where e is the electronic charge, R_T is the total resistance of SET, K_B is the Boltzman constant, and ΔF is the Helmholtz's free energy difference. In the proposed model we use free energy difference between energy levels on each side of the tunnelling barriers and consider electron tunnelling from either side of the barriers. Accordingly, we modified the Helmholtz's free energy difference [9] for electrons tunnel from source to island ΔF_S^+ and from island to source (ΔF_S^-) at the source-island junction as

$$\Delta F_S^{\pm} = \frac{e}{C_\Sigma}\left(\frac{e}{2} \pm \left(V_{DS}\left(C_S + \frac{1}{2}C_G\right) + ne - q_0\right)\right) \mp \Delta E_{island} \tag{2}$$

where q_0 is the background charge, n the external electron number, C_Σ the total capacitance, C_S the source capacitance and C_G the gate capacitance of the SET. Similarly, electrons tunneling from drain to island and from island to drain can be expressed as:

$$\Delta F_D^{\pm} = \frac{e}{C_\Sigma}\left(\frac{e}{2} \pm \left(V_{DS}\left(C_D + \frac{1}{2}C_G\right) + ne - q_0\right)\right) \pm \Delta E_{island} \tag{3}$$

C_D is the drain capacitance of the SET. In Eqs (2) and (3), ΔE_{island} is the external energy of the island induced by external voltage. By substituting Eqs. (2) and (3) into Eq. (1), we can obtain four tunneling rates $\Gamma_S^+, \Gamma_D^+, \Gamma_S^-$, and Γ_D^- for electron tunnels from source to island, from drain to island, from island to source, and from island to drain, respectively. The tunneling rate obtained from the proposed model is non-continuous and has a quadratic relation with V_{GS}, and is different from other proposed model found in the literature [7]

Rather than master equation we expressed the electron current from drain to source as below by considering the two tunnelling junctions in series:

$$e_{DS} = e \cdot \frac{\Gamma_S^+ \cdot \Gamma_D^+}{\Gamma_S^+ + \Gamma_D^+} \tag{4}$$

The electron current flow from source to drain e_{SD} can be expressed in a similar way, and the total current form drain to source can then be written as:

$$I_{DS} = -q\left(e_{DS} + e_{SD}\right) \tag{5}$$

3. Simulation Results

Figure 1 compares the I_{DS}-V_{DS} characteristic of the proposed model and SIMON 2.0 at different temperatures. Good agreement between the two models can be observed.

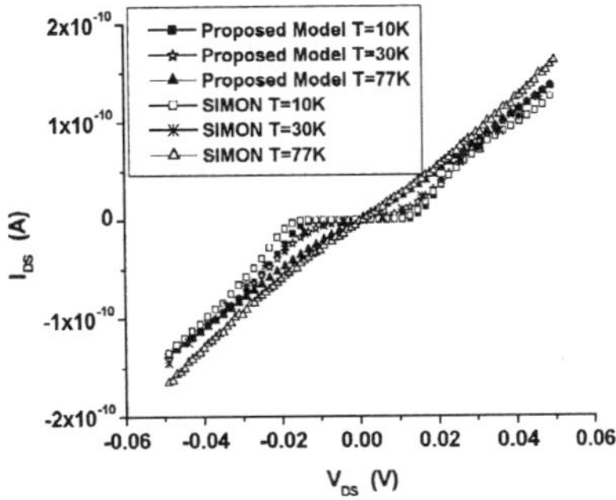

Fig. 1 Comparison of the I_{DS}-V_{DS} characteristic of the proposed model and SIMON 2.0.

Figure 2 shows the I_{DS}-V_{GS} characteristic of the proposed model and SIMON 2.0. The proposed model can give I_{DS}-V_{GS} characteristic quite close to that of SIMON 2.0. Figures 3 (a) and 3 (b) compares the I_{DS}-V_{DS} characteristic obtained respectively from the proposed model and the most-cited analytical model developed by Uchida $et.$ $al.$ [8] at different temperatures (33°K and 77°K). Clearly, the simulation results of the proposed model are closer to those of SIMON 2.0 than the Uchida's model. This is particularly true when the temperature is higher.

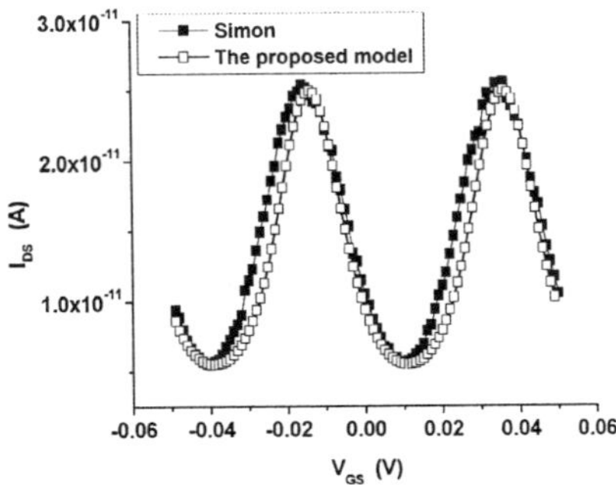

Fig. 2 The simulation results of I_{DS}-V_{GS} characteristic of the proposed model and SIMON 2.0.

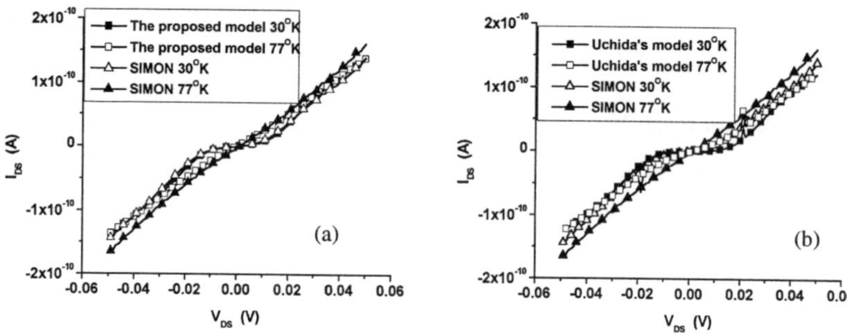

Fig. 3 Comparison of the I_{DS}-V_{DS} characteristic obtained from (a) the proposed model and (b) the analytical model developed by Uchida et. al. [8] at 30°K and 77°K.

4.   Conclusion

In this paper, we presented an analytical single electron transistor model based on the free energy difference between energy levels on each side of the tunnelling barriers and bilateral tunnelling of electrons from either side of the barriers. By comparing our proposed model with Monte Carlo SET simulator SIMON 2.0, excellent match in the simulation results can be obtained. Notably, when compared with other analytical model found in the literature, our proposed model can give less difference with the SIMON 2.0 at higher temperature.

References

1.   K.K.Likharev, *Proceedings of the IEEE*, **87**, 606 (1999)..
2.   C. Wasshuber, H. Kosina, S. Selberherr, *IEEE Trans. on Computer-Aided Design of Integrated Circuits and Systems*, **16**, 937 (1997).
3.   L.R.C. Fonseca, A.N. Korotkov, K.K. Likharev, and A.A. Odintsov, *J. Appl. Phys.*, **78**, 3238 (1995).
4.   R.H. Chen, A.N. Korotkov, and K.K. Likharev, *Appl. Phys. Lett.*, **68**, 1954 (1996).
5.   Y.S. Yu, S.W. Hwang, and D. Ahn, *IEEE Trans. Eelctron Devices*, **46(8)**, 1667 (1999).
6.   Y.S. Yu, J.H. Oh, S.W. Hwang, and D. Ahn, *Electronics Lett.*,**38(16)**, 850 (2002).
7.   X. Wang, and W. Porod, *Superlattices and Microstructures,* **28(5/6)**, 345 (2000).
8.   K. Uchida, K. Matsuzawa, J. Koga, R. Ohba, S. Takagi, and A. Toriumi, *Jpn. J. Appl. Phys.* Part 1, 39(4B), 2321 (2000).
9.   C. Wasshuber, *Computational Single-electronics*, Springer Wien, New York.

LARGE-QUANTITY SYNTHESIS METHOD OF CARBON AND BORON NITRIDE NANOTUBES: MECHANO-THERMAL PROCESS

YING CHEN

Department of Electronic Materials Engineering, Research School of Physical Sciences and Engineering, The Australian National University, Canberra, ACT 0200, AUSTRALIA

This paper reports a new synthesis method for both carbon and boron nitride nanotubes. The mechano-thermal method consists of high-energy mechanical milling and thermal annealing processes. Nanotubes with different tubular structures and sizes have been produced. Advantages of this new method include large quantity production in the order of kilograms, low production costs and controllable process. The metal catalysis and new solid-state formation mechanisms involved are discussed.

1. Introduction

Nanotubular materials including carbon and boron nitride (BN) have excellent mechanical and physical properties compared with their bulk materials. The seamless graphitic cylinders with a high length to diameter ratio (more than 1000) make them as super strong fibers to enforce composite materials [1]. Its special electronic properties varying with diameter and helicity have been used to make nanosized transistors [2]. The smallest cylinders can store huge amount of hydrogen gas due to the large surface area and micro porous structure and thus create a bright future for hydrogen-powdered cars [3]. All these promising industrial applications require obviously a large quantity of nanotube materials in the order of kilograms or above. However, mass production in the order of kg of high quality nanotubes is still difficult. Discovery of new synthesis methods is still crucial for any application in industrial scale. A new mechano-thermal process was discovered by the current author in 1999 [4, 5]. This mechano-thermal process consists of mechanical milling and thermal annealing processes, and can produce nanotube materials in the order of kilograms [6]. In addition, because both milling and heating processes are fully controllable and at relatively low temperatures (less than 1400 °C), it provides an excellent opportunity to investigate formation mechanisms of nanotubes, which have not yet been fully understood. Another advantage of this method is that boron nitride nanotubes can be produced readily. BN nanotubes have similar tubular structure as carbon ones, but with a better thermal conductivity, more stable electronic properties and higher chemical stability [7, 8]. However, BN nanotubes in large quantity are difficult to produce previously. This paper demonstrates that carbon nanotubes have been produced in graphite samples after pre-milling and subsequent annealing treatments. BN nanotubes can be prepared via either ball-milling induced nitridation reaction between pure elemental B powder and ammonia gas [4], or mechanically activated solid-state crystal growth [9, 10]. Finally, new formation mechanisms are discussed.

2. Experimental procedure

High-purity crystalline graphite, boron (B) and BN compound powders were used as the starting materials. Ball milling experiments were performed at room temperature in a vertical planetary ball mill using hardened steel balls with a diameter of 25.4 mm and a stainless steel container. A schematic illustration of ball milling actions is shown in Fig. 1. The milling container was loaded with several grams of starting powders together with four milling balls. A starting pressure of about 300 kPa of ammonia or argon gas was established before each milling experiment. Heat treatments were conducted under a N_2 gas flow at temperatures up to 1400 °C in a tube furnace. Milled and annealed samples were characterized using various analyzing techniques. Crystalline structure of samples was investigated by means of X-ray diffraction analysis (XRD) using Co radiation (λ=0.1789 nm) at room temperature. A Hitachi S4500 field-emission scanning electron microscope (FESEM) is used to imaging nanosized fibers and tubular structures were observed using a Philips EM430 (300kV) transmission electron microscope (TEM). The average metal concentrations were examined by using X-ray energy dispersive spectroscopy (XEDS) in a JEOL (JSM6400) scanning electron microscope.

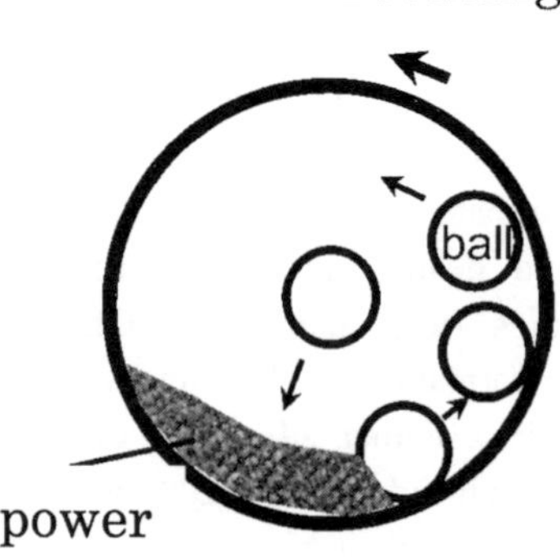

Fig. 1 Schematic representation of milling actions inside milling chamber.

3.   Results

3.1. *Carbon nanotubes*

A graphite sample was first milled for 150 hours and then heated at 1400°C for 6 hours in flowing nitrogen gas. Carbon nanotubes in the web form were found in the annealed sample as shown by a typical TEM micrograph in Fig. 2. The outer diameter of the tubes is less than 20 nm. The inter diameter is in the range of a few nanometers. No metal particle is observed inside these tubes but dark metal particles are observed around. These particles are iron contaminants from steel milling balls and the container. Multi-walled nanotubes containing a particle at the tip were also observed. In these cases, the iron particles should act as catalysts for the nanotube formation, as suggested in other processes [11]. Most nanotubes form sub-micron sized clusters.

No nanotubes were found in the sample directly after milling treatment without subsequent heat treatment. Also, direct annealing of the starting graphite powder without pre-milling treatment does not produce any nanotube. Therefore, the nanotubes should grow only during annealing process from the pre-milled sample with or without the help from the iron particles. It is found that the milled sample has a nanoporous and disordered structure [12]. Because the temperature at which nanotubes are produced is much lower than the melting point of graphite, vapor phase should not be produced. Nanotubes are produced in this special process in a solid state and thus with a new formation mechanism.

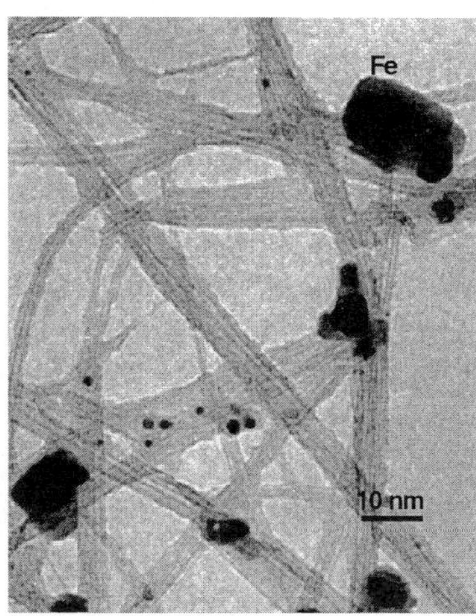

Fig. 2 TEM micrograph of carbon nanotubes produced by mechano-thermal process. The particles with dark contrast are iron.

3.2. *Boron nitride nanotubes*

BN nanotubes can be produced with two different reaction processes.

3.2.1. *Reactive ball milling*

The starting pure boron powder was milled in ammonia gas at a starting pressure of 300 kPa for up to 150 hours. During milling, a huge amount of gas was absorbed onto the fresh surface of the powder created by high-energy ball impacts and even nitridation reaction was induced during extended milling, as indicated by pressure decrease and re-increase inside the milling chamber [4]. Nanocrystalline BN was produced at the end of milling, which serve as seeds for nanotube growth during subsequent heating process. Subsequent heat treatment of the as-milled sample under nitrogen gas flow induced progress of the nitridation reaction at a temperature of 1300 °C for 6 hours. XRD analysis reveals the formation of hexagonal BN phase and small amount of iron boride due to iron contamination from milling media (balls and container). Fine filaments extruding from BN aggregates are observed with FESEM, as shown in Fig. 3. TEM reveals that these filaments are hollow with a tubular, multi-walled structure and a typical high resolution image is shown in Fig. 4.

Fig. 3 FESEM micrograph of BN nanotubes.

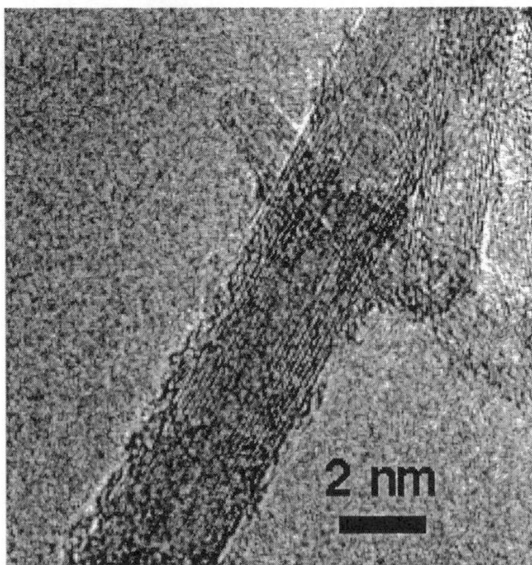

Fig. 4 TEM image of multi-walled BN nanotubes.

3.2.2. *Mechanically activated crystal growth*

The starting material in this case is crystalline boron nitride compound. The BN powder was milled for 140 hours in nitrogen gas and then heated at 1200 $^{\circ}$C for 10 hours. Multi-walled BN nanotubes with an outer diameter of about 11 nm and an inner diameter of 3 nm were produced. Special bamboo-type tubes were observed in the same milled sample after annealing at a higher temperature of 1300 $^{\circ}$C for 10 hours (Fig. 5), revealing a typical catalytic role of metal particles. The BN nanotube formation process in this case is similar to carbon nanotubes because of absence of any nitridation reactions. Therefore, nanotube formation should be the result of nanocrystal growth at low temperature. Detailed mechanism is published in another publication. This result confirms that the nanosized BN particles produced by reactive ball milling (the above case) could serve as seeds.

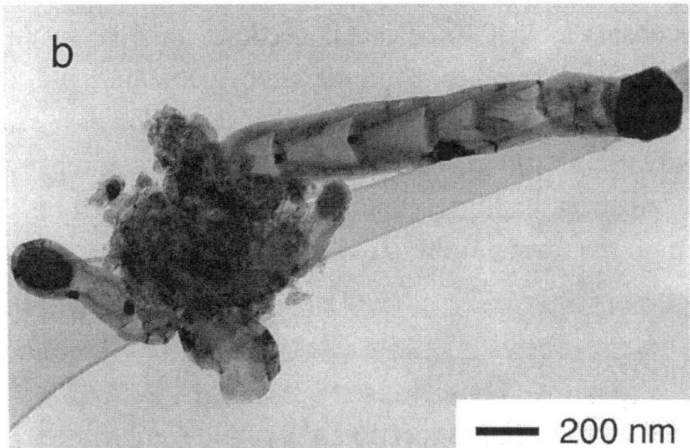

Fig. 5 TEM image of BN nanotubes with a bamboo-type structure and iron particles at tip.

4. Formation mechanisms discussions

4.1. *Nucleation*

Mechanical treatment (high-energy ball milling) has an essential role in the mechano-thermal process. Without milling treatment, direct heat treatment of graphite or BN samples cannot produce any tubular structures. Nanotubes were not observed in any milled samples either. These observations suggest that the milled materials may contain a special nucleation structure. Both XRD and electron microscopy analyses reveal that milled samples have nanocrystalline phases and are highly disordered (or close to amorphous) structure. The nanocrystals are seeds for nucleation and the amorphous phases are mobile atom source for tube growth. Therefore, mechanical treatment actually produces a nanotube precursor.

4.2. *Nanotube growth*

Nanotubes start to grow out from the precursor powders when the milled samples are heated at certain temperatures. This growth process appears to occur with a new diffusion mechanism. First, the growth temperature (close to the annealing temperature at 1400 °C) is significantly lower than the melting points of graphite or BN (above 2000 °C), the nanotubes grow in solid state or similar to crystallization growth. At these low temperatures, surface diffusion should dominate the atomic diffusion and is the main mechanism of nanotube growth [13, 14].

4.3. *Metal catalysts*

It was directly observed that some metal particles involve in the formation of certain nanotubes. For examples, some nanotubes contain metal particles inside or at tips. These particles are known to serve as seeds and also maintain the tube open during growth. In our case, the metal particles, coming from the milling media as contamination, catalyse nanotube growth. Therefore, they are useful rather than harm to the formation process.

5.   Conclusions

Carbon nanotubes have been produced in a graphite sample after pre-milling and subsequent annealing treatments. BN nanotubes can be prepared via either ball-milling induced nitridation reaction between pure elemental B powder and ammonia gas or mechanically activated solid-state crystal growth during subsequent thermal annealing. Metal particles coming from the milling media contamination catalyse nanotube growth. Surface diffusion should dominate the atomic diffusion and is the main mechanism of nanotubes growth in solid state at such low temperatures.

Acknowledgments

The author would like to thank Mr Martin Conway and Ms Jun Yu for the help in conducting milling and annealing experiments, and Professors J.S. Williams, L. T. Chadderton, and Dr J. Fitz Gerald for their valuable discussions. Dr Sally Stowe and staff of the Electron Microscopy Unit are gratefully acknowledged for their support in using various microscopy facilities. This project is partially supported by several Research Grants from the Australian Research Council (Grant No. 8222693, No. F99086 and F00156).

References

1. P.M. Ajayan, O. Stephan, C. Colliex, and D. Trauth, *Science* **265** 1212 (1994).
2. S.J. Tans, ARM Verschueren, and C. Dekker, *Nature,* **393,** 49 (1998).
3. C Liu, YY Fan, M Liu, HT Cong, HM Cheng, MS Dresselhaus, *Science*, **286**, 1127 (1999).
4. Y. Chen, J. Fitz Gerald, J.S. Williams and S. Bulcock, *Chem. Phys. Lett*, **299**, 260 (1999).
5. Y. Chen, J. Fitz Gerald, L.T. Chadderton and L. Chaffron, *J. Metastable and Nanocrystalline Materials*, **2-6** 375 (1999).
6. Y. Chen, M. Conway, J.S. Williams, J. Zou, *J. Materials Research*, **17(8)** 1896 (2002).
7. N.G. Chopra and A. Zettle, *Sol. State. Communication*, **105**, 297 (1998)
8. X. Blasé, A Rubio, S. G. Louie and M.L. Cohen, *Europhys. Lett.* **28**, 18369 (1994).
9. Y. Chen, L.T. Chadderton, J. Fitz Gerald and J.S. Williams, *Appl. Phys. Lett.,* **74**, 2960 (1999).
10. Y. Chen, L.T. Chadderton, J.S. Williams, J. Fitz Gerald, *Materials Science Forum,* **343-346,**. 26 (2000).
11. C. Laurent, E.Flahaut, A. Peigney and A. Rousset, *New J. Chem.* **1229** (1998).
12. Y. Chen, J. Fitz Gerald, L.T. Chadderton and L. Chaffron, *Appl. Phys. Lett.,* **74(19),** 2782 (1999).
13. Y. Chen, LT Chadderton, JS Williams and J. Fitzgerald, *J. Metastable and Nanocrystalline Materials*, **8**, 63 (1999).
14. L. Chadderton and Y. Chen, *Phys. Lett.,* **A263**, 401 (1999).

DETERMINATION OF THE SPECIFIC SURFACE OF γ-AL$_2$O$_3$ USING SMALL-ANGLE X-RAY SCATTERING

C. MAITLAND, C.E. BUCKLEY, G.PAGLIA, J. CONNOLLY.

Department of Applied Physics, Curtin University of Technology, Perth, WA 6845 Australia.

The temperature dependence of the specific surface area, the average pore radii and the pore volume fraction for powdered γ-Al$_2$O$_3$ are determined using small angle X-ray scattering. The methodology for determining the above parameters for a powdered sample is a modification of the recent work of Spalla et al. In this paper we show that the above parameters can be determined from SAXS by using a standard to convert the scattered intensities to an absolute scale (cm^{-1}).

1. Introduction

Gamma alumina (γ-Al$_2$O$_3$) is a transition alumina reported to occur at temperatures between 350 and 1000°C. It is typically formed from an amorphous or boehmite precursor, and has remained present at temperatures as high as 1200°C in the former case. Industrially, it is extremely important. It is used as a catalyst and catalyst support in the automotive and petroleum industries, in structural composites for spacecraft, miniature power supplies, and abrasive and thermal wear coatings [1]. It is important to study the evolution of the porosity with temperature in γ-Al$_2$O$_3$, especially since the porosity is important in determining the overall structure [2].

Recently Spalla et al. [3] derived a method to analyse the small angle X-ray scattering (SAXS) intensity from a porous and granular medium. Their method relies on determining the absolute scattering intensity (cm^{-1}) from a knowledge of the raw count rates, the transmission of the sample, the solid angle, and the thickness of the sample. Our method differs from Spalla et al. [3] in that a standard is used to convert the measured intensity to the absolute scattering intensities, hence a knowledge of the raw count rates and solid angle is not required. Using this method the evolution with temperature of porosity, specific surface area and pore radii is determined for γ-Al$_2$O$_3$ powders.

2. Experimental

Sample Preparation

Powdered γ-Al$_2$O$_3$ samples were produced from the calcination of highly-crystalline hydrogenated boehmite, obtained from the Alumina and Ceramics Laboratory, Malakoff Industries, Arkansas, USA. Each boehmite precursor was calcined at temperatures between 500 and 900°C, at 50° intervals, in air. The heating rate was 5° per minute. Calcination of each sample lasted for seven hours at each temperature.

SAXS

The small-angle intensities were measured with the NanoSTAR SAXS instrument at Curtin University. Scattered photons were counted with a 2-D multiwire detector over a period of three hours. Instrumental settings were as follows; $\lambda = 1.5418$ Å (Cu Kα); sample detector distance of 65 cm and 2θ range between 0.04 and 4.4 degrees. The raw data had the background subtracted and was radially averaged. Resulting intensities after discarding low q data points affected by the beam stop, spanned a q range between $0.01138 - 0.31287$ Å^{-1} ($q = 4\pi\sin\theta/\lambda$).

3. Method of SAXS Analysis

The approach given here for the analysis of small-angle x-ray scattering (SAXS) data from powder samples is the same as that described in [3]. This approach allows one to determine the volume fraction φ of nanometre-sized porosity and specific surface Σ_S of specimens in powder form. This analysis is valid when the size of the powder grains is more than ten times the size of the porosity. Difficulties arise when trying to determine the nano-structure of materials that are in powder form. The cause of these difficulties is that the structure of interest is located within the grains of the powder. This internal structure is the same for each grain. The "envelope" structure however (grain size, shape and packing) will change from sample to sample, regardless of the internal structure. The schematic shown in figure 1 aims to explain such things as the "grain envelope", "material layer" and "solid thickness". These concepts were developed by Spalla *et al.* [3] and are important in understanding how useful information can be derived from the small angle intensity of powder samples.

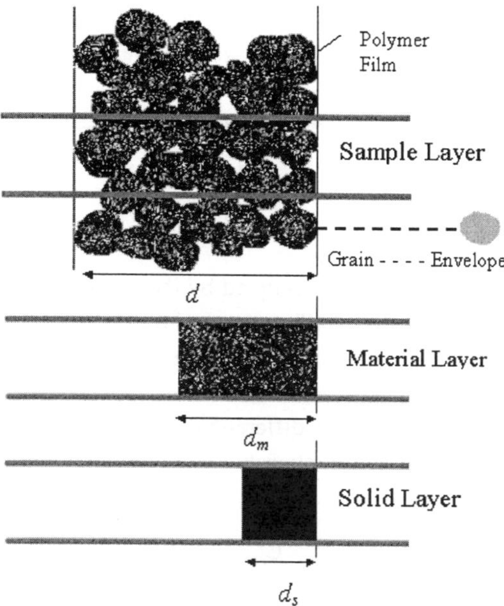

Figure 1: Schematic that shows the typical sample arrangement for a SAXS measurement of powder samples. The powder is sandwiched between two thin polymer films that are less than a millimetre apart. This figure has been adapted from Spalla, Lyonnard and Testard (2003).

A grain envelope is defined as a micrometric sized grain with its outer boundary smoothed of any small scale structure e.g. this structure may be nanometre sized pores. The material layer is the equivalent amount of porous medium (solid + inner pores) that the x-ray beam passes through. The volume of this material layer is equal to that occupied by the envelopes of the grains V_e. Similarly, the solid layer is the equivalent amount of solid through which the X-ray beam passes. The transmission T of the sample is dependent on the solid thickness d_s.

When the interfacial surface area between two phases is proportional to the mass m of one of the phases, the specific surface Σ_s is defined as the interfacial surface area, S per unit mass of this phase [4].

$$\Sigma_s = \frac{S}{m} \tag{1}$$

For a porous solid with air filled pores, the two phases are air and the solid material; S in this case is interfacial surface area between the air and the solid, and m is the mass of the solid. The BET method is a complementary technique for calculating the specific surface of porous materials, however this method cannot be used to measure closed porosity. One usually wishes to determine the specific surface Σ_s and the volume fraction φ of the grains occupied by the nanometre-sized pores. The volume fraction φ is defined as

$$\varphi = \frac{V_p}{V_e} \tag{2}$$

where V_p is volume of the grains occupied by the nanometre sized porosity. We see that regardless of how the powder is packed the quantity φ will not change.

The raw intensity $I(q)$ was converted to the 'measurable' intensity, which is on an absolute scale $I_1(q)$ (cm^{-1}). When the sample comprises of pore and solid phases only, $I_1(q)$ represents the differential scattering cross section per unit volume of solid. The measurable intensity can be determined directly [3];

$$I_1(q) = \frac{I(q)}{C_o \Delta\Omega(q) T d_s t} \tag{3}$$

where T is transmission of the powder sample and d_s the length of solid traversed by the incident beam. This method requires the measurement of the number of incident photons per second C_o, and number of photons per second $I(q) \, / \, t$ scattering into the solid angle $\Delta\Omega(q)$. The NanoStar SAXS instrument is not set up to perform direct measurements. Instead a highly cross-linked polyethylene S-2907 standard supplied from Oak Ridge National Laboratories was used to determine the measurable intensity;

$$I_1(q) = I_{abs,st} \frac{I(q)}{I_{st}} \frac{t_{st}}{t} \frac{d_{st}}{d_s} \frac{T_{st}}{T} \tag{4}$$

where $I_{abs,st}$ is the differential scattering cross section per unit volume of the standard at $q = 0.0227$ Å^{-1}; I_{st} is the number of photons detected in time t_{st} at $q = 0.0227$ Å^{-1} and d_{st} the measured thickness of the standard. The solid thickness d_s is calculated from the measured transmission T, and the linear attenuation coefficient of the solid;

$$d_s = -\frac{\ln(T)}{\mu} \tag{5}$$

Using the Porod law [5] the specific surface can then determined with the following expression [3]

$$\Sigma_s = \frac{1}{\rho_m} \frac{\lim_{q\to\infty}\left[I_1 q^4\right]}{2\pi\Delta\rho^2} = \frac{1}{\rho_m} \frac{K}{2\pi\Delta\rho^2} \qquad (6)$$

where ρ_m is the mass density of the solid and $\Delta\rho^2$ is the scattering length density difference between the pore and solid. The specific surface calculated using equation (6) includes all interfacial surface area between pores and solid, plus that between the surface of the grains and air. K was determined by fitting of constant to plots of $I_1 q^4$ vs q over appropriate ranges of q (see figure 3).

To calculate the pore volume fraction, the scattering from the 'envelopes' of the grains has to be subtracted. The measurable intensity after subtraction of scattering from the outer grain boundary is denoted by I_1^{corr} and is determined from Porod scattering, Pq^{-4} at low q where P is the Porod constant. In the low q region the intensity contribution from the nanometre sized porosity is negligible. The pore volume fraction was then calculated using the following expression [3]

$$\int_0^\infty I_1^{corr} q^2 dq = 2\pi^2 \varphi \Delta\rho^2 \qquad (7)$$

I_1^{corr} is only known over a finite q range, thus to perform this integration one has to extrapolate I_1^{corr}. At low q, I_1^{corr} was estimated by a constant value, and at high q the corrected intensity was estimated using the Porod law. ie. $\lim_{q\to\infty} q^4 I_1^{corr}(q) = $ constant. If we assume a spherical shape for the pores the radius of the pores within the grains can be determined from [3]

$$r = \frac{3\phi}{\Sigma_s (1-\phi)\rho_m} \qquad (8)$$

Please note that the Spalla *et al.* [3] paper has the incorrect equation.

4. Results and Discussion

Preliminary inspection of the scattering curves in figure 2 shows there is structure on two length scales within the sample. This is consistent with the observed micron-sized envelope structure of the grains and nano-sized porosity within the grains. Also evident from the curves in figure 2, is that the temperature at which the sample is calcined clearly effects the nanoporous structure of sample. We see that as the temperature increases, the q value at which the pore scattering becomes significant decreases. This suggests the size of the porosity is increasing with temperature.

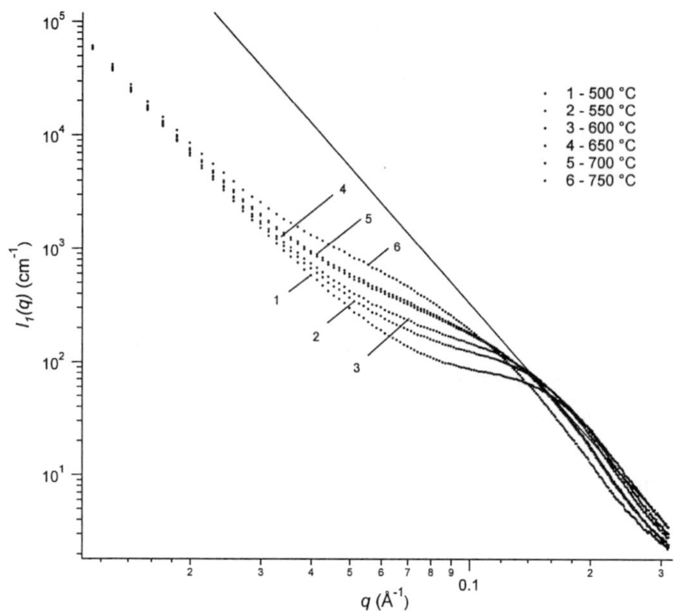

Figure 2: The measurable intensities of the samples that have been calcined at different temperatures. These plots clearly show that the calcination temperature affects the nanostructure. The line shows scattering that is proportional to q^{-4}.

The Porod plots in figure 3 of the measurable intensities from samples calcined at higher temperatures (650 – 750 °C) exhibited a significant increase at high q. In these cases K was calculated from the plateau of the local minimum of $I(q)q^4$. A similar method was used in Schaefer et al. [6] (1995) for the determination of K for scattering curves measured from pores with fractally rough surfaces. The consequence of this is that we are averaging out any short scale structure, which may lead to an underestimation of the actual specific surface. Using this analysis we are essentially probing porosity of no less than about 20 - 23 Å in size. The increase of $I(q)q^4$ at high q suggests that some small porous structure ($r < 1$ nm) still remains in these samples.

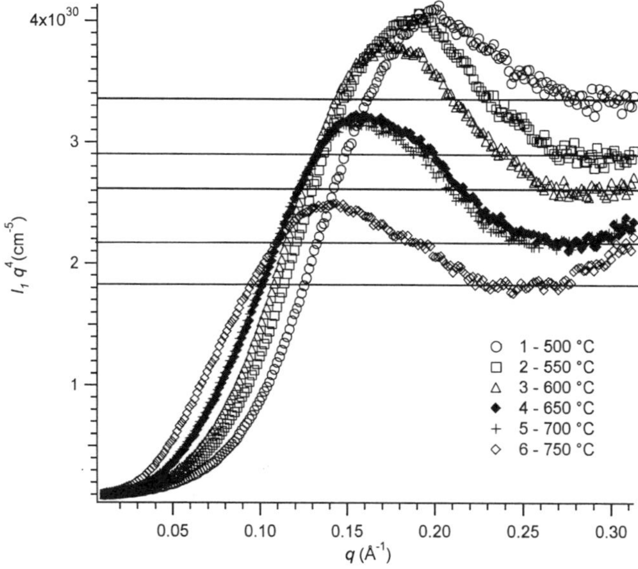

Figure 3: Porod plots of the measurable intensities. The lines show the fitted values obtained for K.

The calculated specific surface, the calculated volume fraction of the nanoporosity within the grains and the average pore radii for each of the samples is shown in table 1. The results shown in table 1 suggest that the pores may be fusing together during calcination, and hence increasing the calcination temperature promotes a greater fusion of pores. As the pores fuse their specific surface decreases, whilst their total volume remains constant.

Table 1: The volume fraction φ and specific surface Σ_S of the pores in each sample.

Temperature (°C)	φ	Σ_S (m^{2}g^{-1})	r (nm)
500	0.172 +/- 0.017	157 +/- 8	1.1 +/ 0.1
550	0.180 +/- 0.018	135 +/- 7	1.3 +/ 0.1
600	0.184 +/- 0.018	122 +/- 6	1.5 +/ 0.1
650	0.180 +/- 0.018	101 +/- 5	1.8 +/ 0.2
700	0.182 +/- 0.018	101 +/- 5	1.8 +/ 0.2
750	0.182 +/- 0.018	85 +/- 4	2.1 +/ 0.2

5. Conclusion

A standard can be used to convert measured SAXS data to an absolute scale (cm^{-1}), which allows a modification of Spalla *et al.* [3] method to be employed for the determination of the temperature evolution of pore volume fraction, specific surface area and average pore size within the grains of γ-Al_2O_3 powders. Our results show that the pore radii increases with temperature with a concomitant decrease in specific surface area, and within experimental error the pore volume fraction remains constant.

Acknowledgments

C.B. acknowledges the financial support of the Australian Research Council for REIF grant R00107962 2001, which enabled the SAXS studies to be undertaken. GP would like to thank the Australian government for the granting of an Australian Post Graduate Award with Stipend (APAWS) and the Australian Institute of Nuclear Science and Engineering (AINSE) for an AINSE Post-Graduate Research Award (APGRA). J.C. is supported by a fellowship from the Australian Nuclear Science and Technology Organisation and the Western Australian Petroleum Research Centre.

References

1. G. Paglia, C.E. Buckley, A. Rohl, B.A. Hunter, R.D. Hart, J.V. Hanna and T. Byrne, *Accepted in Phys. Rev. B.* (2003).
2. G. Paglia, C.E. Buckley, A. Rohl, R.D. Hart, K. Winter, A.J. Studer, B.A. Hunter and J.V. Hanna, *Submitted to J. of Chem. Mater.* (2003).
3. O. Spalla, S. Lyonnard and F. Testard, *J. Appl. Cryst.*, **36**, 338 (2003).
4. D.A. Mcnaught and A. Wilkinson, *IUPAC Compendium of Chemical Terminology*, 1997.
5. G. Porod. *Small Angle X-ray Scattering*, eds. O. Glatter and O. Kratky, Academic Press, New York, 1982. p. 17.
6. D.W. Schaefer, R.K. Brow, T.R. Olivier, G. Beaucage, L. Hrubesh and J.S. Lin, *Modern Aspects of Small-Angle Scattering*, ed. H. Brumberger, Kluwer Academic Publishers, Netherlands, 1995 p. 299.

A MODIFIED SCANNING ELECTRON MICROSCOPE AS A NANOLITHOGRAPHY TOOL

GRAHAME ROSOLEN

CSIRO Telecommunications & Industrial Physics
Vimiera Road, Marsfield, NSW 2121, Australia

Electron beam lithography is an important technique for fabricating devices with dimensions on the nanometre scale. The high cost of commercial electron beam lithography equipment has motivated the development of an electron beam lithography instrument based around the electron optics, sample stage and vacuum system of a scanning electron microscope (SEM). The instrument is designed for direct-write and mosaic electron beam lithography and has been used to fabricate nanoscale devices. A dedicated computer controlled pattern generator and pattern alignment system has been developed and interfaced to the deflection coils, electron detector and beam blanking of the SEM. This provides digital control of the electron beam deflection coils for exposing patterns and also enables images to be acquired for use in pattern alignment prior to exposure. A novel image correlation technique is used to align the exposed patterns with structures already on the sample. This is useful for exposing multiple layered structures that require precise alignment between successive layers. A focus compensation algorithm has been developed to ensure that the beam is at the optimum focus during exposure. The electron beam energy may be set up to a maximum of 40 kV with beam currents from 10 pA to 1 nA. The field size may be continuously varied and is typically in the range 100 μm to 600 μm. Devices have been written on blank samples as well as accurately aligned with features already patterned on the sample. The instrument has been used to write devices which are contained within a single writing field such as a nanowire, optical detectors and high electron mobility transistors. Patterns which cover multiple adjacent fields have also been written to fabricate diffraction gratings and optically variable devices.

1. Introduction

The high cost and complexity of commercial high resolution lithography instruments [1-3] has motivated the development of an electron beam lithography instrument based around the electron optics and sample stage of a scanning electron microscope (SEM) [4, 5]. This configuration reduces the capital and operating costs of the lithography equipment and the infrastructure requirements. Furthermore, the instrument design enables the pattern preparation, lithography and subsequent analysis of the fabricated structures all to be carried out with the instrument. The instrument has been used in a range of direct-write and mosaic lithography applications. In direct-write lithography the patterns exposed with the electron beam must be accurately aligned with other features already on the sample. In mosaic lithography the sample is a featureless substrate and the electron beam is used to expose multiple fields that are tiled together to produce the complete pattern.

2. Instrumentation

The main components of the instrument include the electron optics, sample stage, vacuum system, pattern alignment system and pattern exposure system. The electron optics, vacuum system and sample stage of a LEO 440 SEM have been integrated with a pattern alignment and pattern exposure system to provide the lithography functions. The operator console integrates the separate SEM and lithography computers. The SEM computer is used to configure the electron optics, control the vacuum system and to provide for interactive imaging. The lithography computer is used to prepare the pattern data files, sequence the lithography functions and to control the dedicated lithography electronics.

The electron optics comprises a tungsten or LaB_6 electron source followed by a pair of condenser lenses, deflection coils, electrostatic blanking plates and an objective lens. A secondary electron detector is used for imaging. The maximum beam energy is 40 kV and beam currents typically range from 10 pA to 1 nA. The exposing field size varies from around 100 μm up to 600 μm. Low beam currents are used for fine linewidth lithography as the electron probe size reduces with the beam current [6]. High beam currents are useful for rapidly exposing larger features where the linewidth requirements are less critical. Smaller field sizes are typically used in conjunction with low beam currents and larger field sizes are used with high beam currents.

The standard sample stage of the SEM is used with the addition of a sample holder plate which is designed for mounting semiconductor wafers or disks of 2, 3 or 4 inch diameter and can also accommodate 4 inch mask plates. The sample stage is motorised in the x, y, z, tilt and rotate axes with horizontal travel of 100 mm x 120 mm. The z axis motor is used to set the working distance which is typically 15 mm. The stage is able to achieve a linear positioning accuracy of about 5 μm. In applications where this accuracy is not sufficient for placement of the lithography patterns, a pattern alignment system has been developed. The pattern alignment system uses an image of the features that are already on the sample to align the beam deflection. These alignment features are typically components of the devices defined using previous electron beam lithography or optical lithography. Once the sample stage has positioned the device under the beam a low resolution image is acquired of the device. This image is compared with a reference image of a device of the same geometry in order to determine the positional error. A hierarchical cross-correlation technique is used to determine the offset between the reference image and the acquired image [7]. This offset is applied to the beam deflection coils and a new image is acquired.

This image is compared again with the reference image to determine if any further adjustment is necessary. Once the offset error is below the required limit the pattern is exposed. The correlation technique has the flexibility to be used to locate features at different magnifications with an accuracy of better than 0.1% of the image field size. As the placement accuracy typically scales with the size of the device, the technique may be used to expose a wide range of device sizes and alignment tolerances.

In order to ensure that the beam is correctly focused at all points on the sample, a focus compensation algorithm has been developed. This algorithm uses the focus values obtained at three positions on the sample to determine the equation of the plane which describes the motion of the sample. This plane fitting technique is used to interpolate the correct focus value at each position. The focus values are automatically used to directly adjust the objective lens current as the stage is moved so that the correct focus is maintained at all points on the sample.

The pattern generator consists of the dual processor architecture of a computer and digital signal processor (DSP) system as well as associated analog and digital electronics. The pattern generator reads the pattern data from the computer and generates the signals which drive the beam blanking and deflection system of the electron optics. It also interfaces with the electron detector to acquire image data. The computer is interfaced directly to the DSP which has its own dedicated software for controlling the beam deflection system, electron detector signal and sequencing of the beam blanking. This software is used to control the optics for both acquiring images and exposing patterns. The DSP provides full digital control of the beam so that it may be used in a variety of scanning modes and digitally repositioned anywhere within the field of view. The rate at which the beam is deflected is controlled by the DSP to ensure that accurate doses may be delivered to the sample during exposure. The interface to the electron column is provided by a pair of digital to analog converters (DACs) for the x and y deflection coils and an analog to digital converter (ADC) for the electron detector together with low noise amplifiers and filtering electronics.

The format and syntax of the lithography data files have been designed to provide flexible control of the lithography tasks and to facilitate conversion of the data required for input to the lithography exposing software. The lithography patterns files are usually computer aided design (CAD) files or image files which describe the features to be fabricated. A range of software programs have been acquired or developed to generate the lithography data files needed to expose the pattern features. The lithography files contain information which describes the stage position for the exposure, the coordinates of the shapes that must be exposed and the exposing parameters. The exposing parameters include the beam current, beam velocity and number of overwrites of the beam, which together determine the exposing dose. In the case of direct-write lithography the data files also include a link to the relevant reference images that are used for pattern alignment.

Once the electron optics has been configured the instrument is designed to iterate through the programmed lithography sequence and to turn off at the completion of the lithography tasks. The lithography sequence comprises moving the stage to the required location, acquiring an image, invoking the pattern alignment routines to accurately position the field of view, and then exposing the stored patterns.

3. Direct-Write Electron Beam Lithography

In direct-write lithography applications the patterns exposed with the electron beam must be accurately aligned with features already on the sample. These features may have been patterned previously by electron beam lithography or optical lithography. The advantage of combining optical lithography with direct-write electron beam lithography is that optical lithography is a high throughput process, because the entire field is exposed simultaneously. Optical lithography is ideal for rapidly exposing those features which do not require fine linewidths. Electron beam lithography is a slower serial process and is ideally suited to fabricating the critical elements of the pattern where fine linewidth features are required.

The instrument has been used to fabricate high electron mobility transistors (HEMTs) [8] for monolithic microwave integrated circuits (MMICs). The dimensions of the gate electrode and its placement in the narrow gap between the source and drain electrodes has a significant impact on the performance of the HEMT [9]. An example of a partially fabricated HEMT with four gate electrodes is shown in Figure 1. The source and drain electrodes were patterned using optical lithography. The instrument was used to locate the device and accurately expose the gate electrodes in the middle of the 4 μm gap between the source and drain electrodes. The shape of the gate electrodes has a significant influence on the high frequency performance of the HEMTs. The gate profile shown in Figure 2 is an SEM image of a cross-section through the exposed electron beam resist after it has been developed.

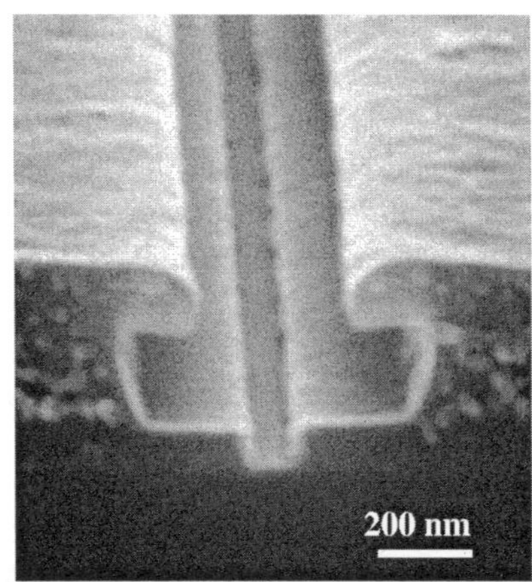

Figure 1. High Electron Mobility Transistor with four gate electrodes.

Figure 2. Resist cross-section showing a 90 nm opening for the gate electrode.

The electron beam resist is used as a mask for etching the substrate prior to forming the gate electrode by evaporating metal over the sample and removing the resist. Three resist layers of different electron sensitivity have been used to produce a T shaped profile [10]. This profile comprises a narrow linewidth at the base, a large cross-sectional area above and a region of resist that overhangs to facilitate cleanly removing the resist and excess metal. In this example the resist opening in the bottom layer is 90 nm. A variety of different HEMT geometries have been written with the instrument ranging from devices with only one gate electrodes up to 16 gate electrodes and with electrode widths covering the range 7 µm to 100 µm.

4. Mosaic Electron Beam Lithography

The instrument has also been used for single layer mosaic lithography which does not require registration with individual features. In mosaic lithography multiple writing fields must be tiled together to produce the complete pattern. This type of lithography has been used to expose patterns to fabricate optically variable devices. These devices exhibit effects that are visible to the naked eye [11]. The instrument is able to fabricate these devices as it has the capability to write patterns covering several millimeters and to expose features with linewidths commensurate with the wavelength of visible light.

A binary switch optically variable device (OVD) is the result of interleaving two different optical patterns in the same physical area of the sample. Each pattern is designed to produce a maximum reflection when viewed from certain angles. The patterns are chosen so that at the maxima angles for one pattern correspond to the minima for the other pattern. This produces the effect when viewing the pattern with the naked eye of switching between the respective images as the viewing angle changes. The device is produced by interleaving pixels from an image of the letter G and an image of the letter R. The pixels corresponding to the letter G are designed to be viewed at right angles to the pixels from the letter R. The respective images visible at the same location on the sample when observed with different viewing angles are shown in Figure 3a and Figure 3b.

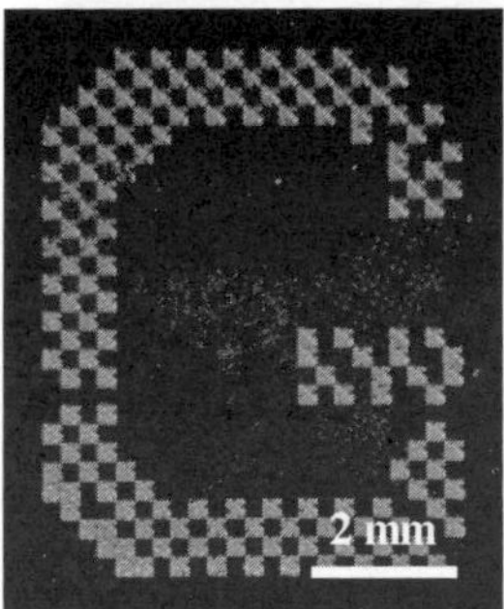

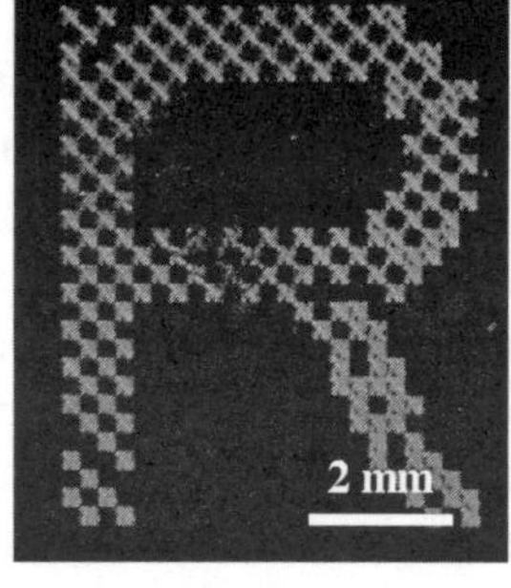

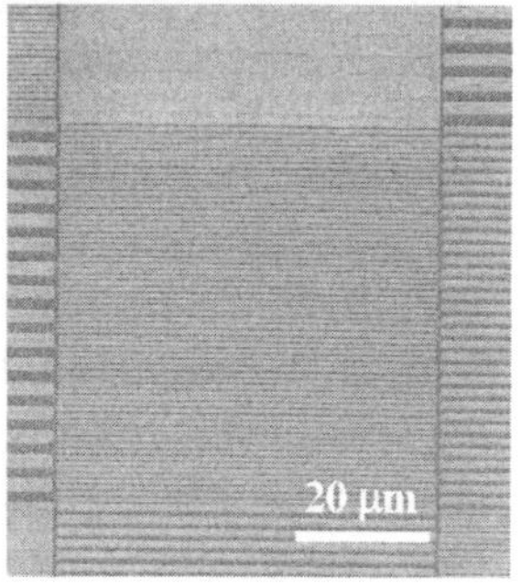

Figure 3a. OVD viewed from the left.

Figure 3b. OVD viewed from the top.

Figure 4. Diffraction gratings for OVD pixels.

Each pixel element is a series of diffraction gratings. The orientation, curvature, width and spacing of the grooves that form each grating is chosen to produce the desired reflection effects. The gratings for the respective images were aligned at right angles to each other as shown in Figure 4. This results in the optical switching effect that the letter which is visible switches from a letter G to a letter R as the sample is rotated. Each pixel is constructed from a 4 x 4 array of gratings. The individual gratings are 56 µm x 56 µm and the number of lines varied from 16 to 64 per grating. This corresponds to a linewidth and spacing of 1.75 µm for the sparse gratings and 0.44 µm for the higher line density gratings. The sample stage was used to step between pixels with a pixel separation of 250 µm. The entire pattern covers an area of 5.7 mm x 7.0 mm. The grating structures were written with a beam energy of 40 kV and beam current of 200 pA. The substrate was a silicon wafers coated with 400 nm layer of polymethylmethacrylate (PMMA) resists [12].

Another type of optically variable device exhibits an optical switching effect with a grayscale image. In this case the grayscale image is produced by a variable density distribution of microdots. Each microdot is 1.25 µm x 1.25 µm and the density of microdots is chosen to correspond to the grayscale value of the pixel. The lithography data files are used to expose the pattern on a reflective substrate coated with a 500 nm high layer of EBR9 electron beam resist. When the sample is developed the grayscale image is visible to the naked eye and has the property that the grayscale inverts depending on the viewing angle. The image in Figure 5 shows the arrangement of the microdots and Figure 6 shows the entire grayscale image when viewed at normal incidence. The image is 5 mm x 5 mm and contains microdots with a printing resolution that corresponds to 20320 dots per inch.

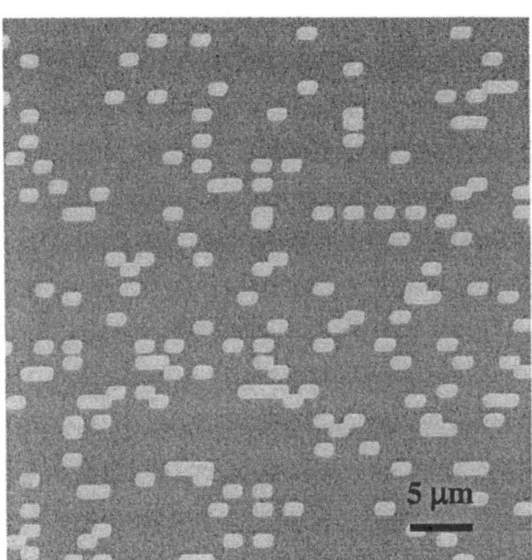

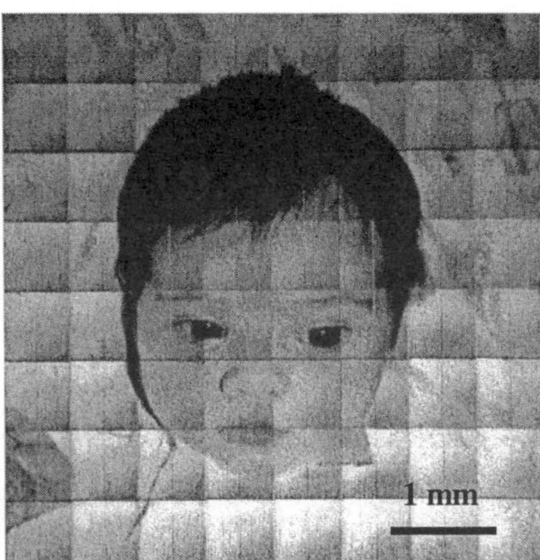

Figure 5. Microdot distribution for grayscale pixels.

Figure 6. Grayscale image viewed at normal incidence.

The original image is 512 x 512 pixels. These image pixels are divided up into groups of 64 x 64 pixels corresponding to a single field. Within each field a pixel is represented by a subfield which is an 8 x 8 array of microdots. These fields are 625 µm x 625 µm and are tiled on an 8 x 8 array to produce the complete image. The beam deflection is used to step through the exposure of each of the subfields and the stage is used to move between subsequent fields. The pattern was exposed with an electron beam energy of 40 kV and a beam current of 1 nA.

128

5. Conclusion

A versatile electron beam lithography instrument has been developed and applied to a variety of direct-write and for mosaic electron beam lithography applications. The instrument has been used to expose fine linewidth structures accurately aligned with features already on the sample. It has also been used to expose patterns to produce optically variable devices. The electron microscopy functions of the instrument are retained and when used together with the pattern location capability of the instrument provide an invaluable tool for inspecting the results of the lithography.

6. References

1. Y. Nakagawa, T. Komagata, Y. Kawase and N. Gotoh, *JEOL News* **38** 1, 32 (2003).

2. F. Abboud, M. Gesley, D. Colby, K. Comendent, R. Dean, W.Eckes, D. McClure, H. Pearce-Percy, R.Prior and S. Watson, *J. Vac Sci. Technol. B* **10** (6), 2734 (1992).

3. K. Okamoto, K. Suzuki, H.C. Pfeiffer and M. Sogard, *Solid State Technology,* **5**, 118 (2000).

4. G.C. Rosolen, *Applied Surface Science* **144-145**, 467 (1999).

5. G. C. Rosolen, *Proc. SPIE* **4404,** 238 (2001).

6. T. Chisholm *J. Vac. Sci. Technol. B* **6** (6), 2066 (1988).

7. G.C. Rosolen and W.D. King, *Scanning* **20** (7), 459 (1998).

8. R. Grundbacher, I. Adesida, Y.C. Kao and A.A. Ketterson, *J. Vac. Sci. Technol. B* **15** (1), 49 (1997).

9. H. Fukui, , *Bell. Syst. Tech. J.* **58** (3), 771 (1979).

10. A.S. Wakita, C.Y. Su, H. Rohdin, H.Y. Lee, J. Seeger and V.M. Robbins, *J. Vac. Sci. Technol. B* **13** (6), 2725 (1995).

11. R.A. Lee, D.F. Lynch, I.J. Wilson, *Optica Acta,* **32**, 573 (1985).

12. M. Hatzakis, *J. Vac. Sci. Technol. B* **12** (6), 1276 (1975).

OPTICAL AND STRUCTURAL INVESTIGATION OF SURFACE-PASSIVATED LEAD SULFIDE QUANTUM DOTS

MARK FERNÉE, ANDREW WATT, JAMIE WARNER, NORMAN HECKENBERG, HALINA RUBINSZTEIN-DUNLOP

Centre for Quantum ComputerTechnology, School of Physical Sciences, The University of Queensland, Queensland 4072, Australia.

JAMIE RICHES

Centre for Microscopy and Microanalysis, The University of Queensland, Queensland 4072, Australia.

Surface passivation of PbS nanocrystals by the introduction of CdS precursors is investigated. The role of CdS in the surface passivation of PbS nanocrystals is uncertain, as the crystalline structure of CdS and PbS are different, which should preclude effective epitaxial growth. Absorption spectroscopy is used to show that the CdS precursors strongly interact with the PbS nanocrystal surface. Electron microscopy reveals that the introduction of CdS precursors results in an increased particle size, consistent with overcoating. However, we also find the process to be highly non-uniform.

Colloidal nanocrystal (NC) quantum dots are nanometer sized semiconductor crystals in which at least one of the intrinsic charge carriers undergoes some degree of quantum confinement. In NCs this is usually seen as a structured absorption spectrum and a blue shift of the absorption band-edge with respect to the bulk semiconductor band-edge. If the NCs are small enough, they can enter the regime of strong quantum confinement, where the confinement-induced energy is larger than all other contributions to the charge-carrier energy. In this regime the intrinsic charge-carriers can be readily described using a simple particle-in-a-box model. It has recently been shown that PbS NCs can be made small enough to operate in this regime [1].

Surface interactions have a dramatic effect on the charge-carrier dynamics of small NCs and cannot be neglected [5]. It is usual practice to modify the surface in some way in order to reduce or prevent the charge-carriers interacting with the surface. We have recently reported a method for surface passivation of small PbS NCs leading to bright photo luminescent emission by introducing CdS precursors [3]. The exact surface passivation mechanism is uncertain due to the crystal structure difference between PbS and CdS crystal structures. However, an epitaxial growth mode is at least possible in one group of crystal directions [4]. Here, we use both optical and TEM studies to show that the introduction of CdS precursors results in a modification of the PbS NC surface and growth in the average particle size. We also show evidence for epitaxial growth.

130

The surface passivation process involves the aqueous synthesis of PbS NCs [5] with a two-fold molar excess of S^{2-} ions, followed by the introduction of Cd^{2+} ions at acid pH (usually pH 4 to 5) [3]. In figure 1 we show the effect on the PbS NC absorption spectrum following the addition of Cd^{2+} ions. While there is a considerable change in the absorption spectrum over time, we do not find any indication of CdS NC growth [3]. The surface passivation procedure terminates the evolution after 20 minutes by raising the pH of the solution to alkaline pH (approximately pH 11). However, if the solution is left for longer, we find that the peaks in the absorption spectrum gradually blue shift and broaden, while new a peak emerges near 6 eV. The emergent peak near 6 eV is due to the presence of Pb^{2+} ions in solution [6,7]. Therefore, the addition of CdS precursors causes a partial dissolution of the PbS NCs, indicating an interaction with and modification of the PbS NC surface.

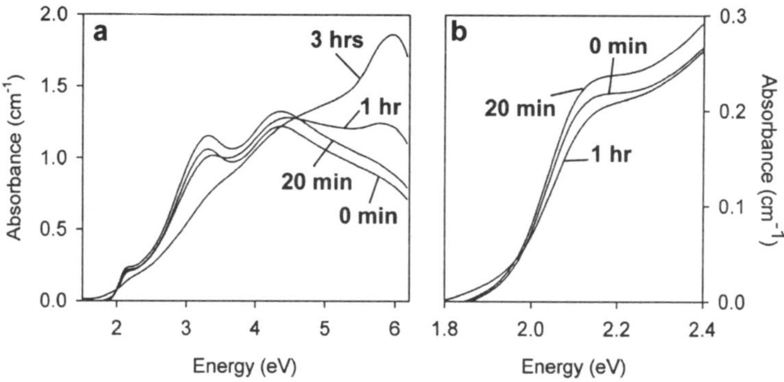

Figure 1. (a) The absorption of a PbS NC solution containing excess S^{2-} ions at different times after the addition of Cd^{2+} ions at a Pb:Cd ratio of 1:1. (b) An expanded view of the effect around the PbS NC absorption band-edge.

The effect of the introduction of CdS precursors is further characterized using transmission electron microscopy (TEM). In figure 2 we present particle size data, from a sample of 485 particles, for both a pure PbS NC solution and the same PbS NC solution after the introduction of CdS precursors (Pb:Cd ratio of 1:2). Here we see that the average size of the NCs is increased by approximately 1 nm after the CdS precursor addition. Furthermore the distribution is broadened towards larger sized NCs, indicating that the CdS precursor treatment is non-uniform.

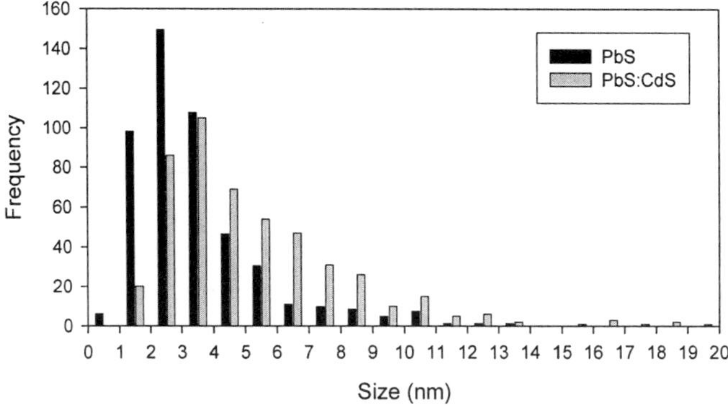

Figure 2. Histogram of particle sizes comparing PbS NCs before and after treatment with CdS precursors (PbS:CdS ratio of 1:2).

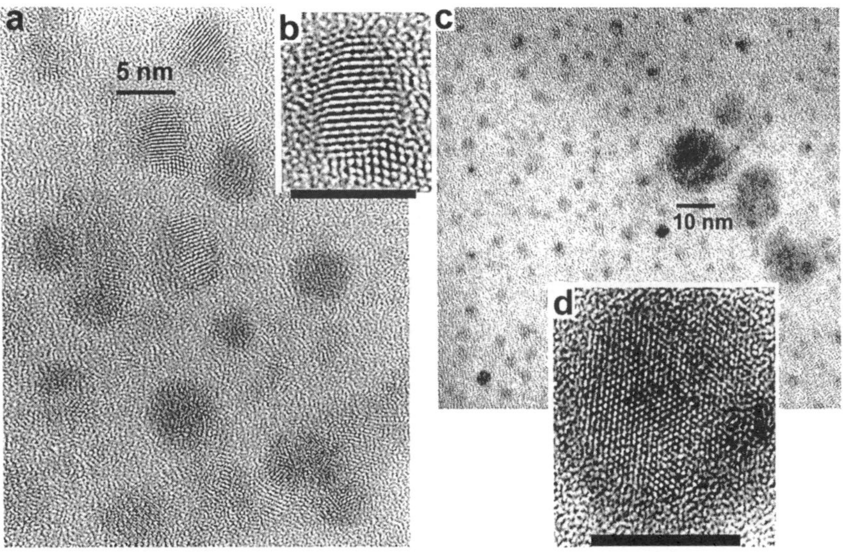

Figure 3. (a) HRTEM image of PbS NCs that have been treated with CdS precursors with a Pb:Cd ratio of 1:1. (b) An expanded view of a polycrystalline nanocrystal providing evidence for epitaxial crystal growth. (5 nm scale bar included) (c) A HRTEM image of PbS NCs that have been treated with CdS precursors with a Pb:Cd ratio of 1:2 and rapid injection of Cd2+ ions. (d) An expanded view of a large nanocrystal that appears nearly mono-crystalline. (10 nm scale bar included)

High resolution TEM (HRTEM) images of PbS NCs after the addition of CdS precursors are shown in figure 3. Nanocrystals of approximately 5 nm are observed in figure 3(a), where a Pb:Cd ratio of 1:1 was used. More importantly, we find that many of the nanocrystals are poly-crystalline. In contrast, for pure

PbS NCs, only mono-crystalline images were observed. The observation of poly-crystalline NCs, such as that shown in figure 3(b), suggests a possible epitaxial growth mode. In figure 3(c) we show another HRTEM image of CdS treated PbS NCs created using a slight variation in the procedure (a Pb:Cd ratio 1:2, rapid injection of Cd^{2+} ions and S^{2-} ions introduced using a Na_2S solution [3]). Here we see that most of the nanocrystals are approximately 3 nm in size with a few significantly larger NCs present. The smaller NCs are primarily PbS NCs, as determined using energy dispersive X-ray spectroscopy (EDX). However, the larger NCs are composite structures containing both lead and cadmium. In general, no pure CdS NCs could be found using EDX. In figure 3(d) we show a larger image of one of the large NCs (10 nm), which has what appears to be a mono-crystalline structure. A number of large mono-crystalline or nearly mono-crystalline NCs were found in this sample. EDX indicates the presence of both cadmium and lead in these NCs. In general the images 3(a) and 3 (c) indicate that the surface passivation is a rather non-uniform process.

We have shown using both optical absorption and HRTEM studies that the introduction of CdS precursors into a colloidal solution of PbS NCs results in strong modification of the PbS NC surface. TEM studies reveal an average increase in NC size after surface passivation, consistent with an overcoating process. Furthermore, large mono-crystalline and poly-crystalline NC images suggest that epitaxial growth may be possible. However, the TEM studies reveal that the surface passivation process is highly non-uniform.

Acknowledgments

MF was funded by the Australian Research Council. We wish to thank John Barry for his help with the electron microscopy.

References

1. Z. Hens, D. Vanmaekelbergh, E. Stoffels, H. van Kempen, *Phys. Rev. Lett.*, **88**, 236803 (2002).
2. Y. Cao and U. Banin, *Angew. Chem. Int. Ed.* **38**, 3692 (1999).
3. M. J. Fernée, *et al.*, *Nanotechnology*, **14**, 991 (2003).
4. S. Watanabe and Y. Mita, *Solid-State Elect.* **15**, 5 (1972).
5. M. T. Nenadovic, M. I. Comor, V. Vasic, O. I. Micic, *J. Phys. Chem.* **94**, 6390 (1990).
6. I. Moriguchi I., *et al.*, *J. Chem. Soc., Faraday Trans.* **94**, 2199 (1998).
7. S. Gallardo, M. Gutierrez, A. Henglein, E. Janata, *Ber. Bun. Phys. Chem.* **93**, 1080 (1989).

TOWARDS MOLECULAR ELECTRONICS: CONDUCTION OF SINGLE MOLECULES

KARL-HEINZ MÜLLER

CSIRO, Telecommunications and Industrial Physics
Sydney, NSW 2070, Australia

We use the TRANSIESTA code to calculate the electrical conduction of single molecules of varying lengths embedded between two gold electrodes. The calculation uses the non-equilibrium Green's function technique in combination with the density functional theory to determine self-consistently the current through a molecule under external bias voltage. We determine the enhancement factors of the electrical resistance of polyene-dithiol, polyphenyl-dithiol and alkane-dithiol of different lengths and discuss the transmission coefficients.

1. Introduction

Molecular electronics, a new emerging science area in the field of nanotechnology, is seen as a possible replacement for silicon device-technology in the next decade [1]. While the advantages of future molecular electronics are undeniable, little is known about the electrical properties of molecules and metal-molecule interfaces [2]. One of the principal experimental difficulties is the formation of electrical contacts between a molecule and the macroscopic world. Predicting the electrical properties of the molecules theoretically is also a complex problem.

Very recently excellent theoretical progress has been made by describing the electrical conduction of single molecules placed between two metal electrodes using first-principle quantum mechanical calculations [3-5]. K. Stokbro et al. [6] have developed the TRANSIESTA code which interfaces the SIESTA electronic structure package for molecules in such a way that the density matrix of the system can be calculated self-consistently when the system is not in equilibrium, i.e. subjected to an external bias voltage. In this *ab initio* approach the charge transfer between molecule and electrodes as well as the voltage drop along molecule and electrode surfaces can be calculated in a parameter free, self-consistent manner.

2. The non-equilibrium Green's function model

The electrical current I through a molecule can be expressed as

$$I = \frac{2e}{h} \int_{-\infty}^{\infty} dE \; T(E,V) \, [\, f(E-\mu_L) - f(E-\mu_R) \,] \tag{1}$$

where f is the Fermi-Dirac distribution characterizing the left (L) and right (R) electrodes, μ_L and μ_R are the chemical potentials of the metal electrodes and V is the applied voltage where $V = (\mu_L - \mu_R)/e$. The transmission coefficient T(E,V) for an electron of incident energy E is given by [6]

$$T(E,V) = Tr\,[\, \mathrm{Im}\,\Sigma_L(E) \; G^+(E,V) \; \mathrm{Im}\,\Sigma_R(E) \; G(E,V) \,] \,. \tag{2}$$

Σ_L and Σ_R are the bulk self-energies used to map the infinite left and right electrodes onto finite ones [7]. The Green's function G(E,V) of the system is evaluated by inverting the finite matrix

$$\begin{pmatrix} H_L + \Sigma_L & V_L & 0 \\ V_L^+ & H_C & V_R \\ 0 & V_R^+ & H_R + \Sigma_R \end{pmatrix} . \tag{3}$$

The Hamiltonian H_C describes the contact region, i.e. the "extended molecule". The "extended molecule" consists of the molecule itself as well as of that part of the left and right electrodes where all the screening takes place, i.e. the first two atomic layers of each metal electrode. H_L and H_R are the Hamiltonians of the finite electrodes (third and fourth atomic layer). V_L and V_R are the interactions between the contact region and the finite electrodes and $\Sigma_{L/R} = V_{L/R}\, g^{L/R}\, V^+_{L/R}$, where $g^{L/R}$ are the Green's functions of the isolated semi-infinite leads. For the Hamiltonians H_C, H_L and H_R, single-electron Hamiltonians are chosen where the effective electron potentials are expressed in terms of the density functional theory containing both Hartree and exchange-correlation potentials. The electron density is obtained by employing non-equilibrium Green's function techniques [6,7] and the calculation is performed self-consistently without any adjustable parameters.

3. Results and discussion

We have calculated, using TRANSIESTA [5,6], the zero voltage resistance and transmission coefficients T(E,V=0) of several dithiol molecules of varying length embedded between Au(111) electrodes. The molecules are polyene-dithiol [-S-(CH=CH)$_{n/2}$-S-] (n=2,4,6,8), abbreviated as PE-n, polyphenyl-dithiol [-S-(Ph)$_n$-S-] (n=1,2), abbreviated as PP-n and alkane-dithiol [-S-(CH$_2$)n-S-] (n=2,4,6,8), abbreviated as A-n.

Fig. 1 shows the calculated resistance R versus the length d_{S-S} of the molecules, defined as the distance between the two sulphur atoms. We find that the resistance of these molecules increases exponentially with length, i.e. $R \sim \exp(\beta\, d_{S-S})$. For the enhancement parameters β we find the values β (PE)=1.9 nm^{-1}, β (PP)=3.5 nm^{-1} and β (A)=7.2 nm^{-1}. These values agree well with experimental data quoted in the literature [8, 9]. Using the WKB method one finds the large value of $\beta = 23$ nm^{-1} for two Au electrodes that are separated by a vacuum gap.

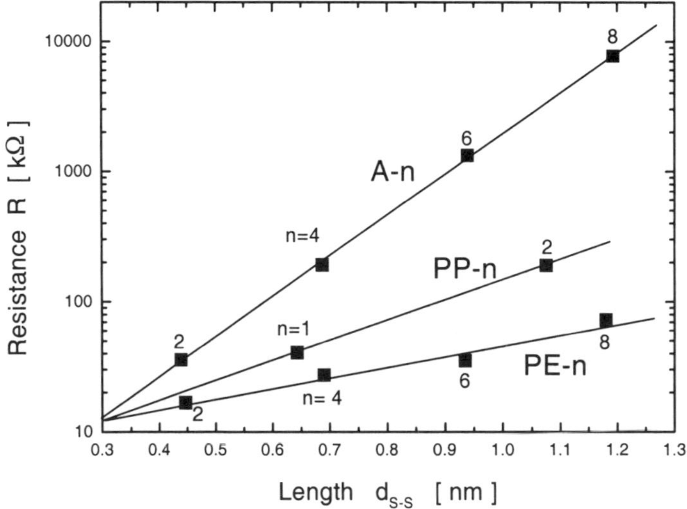

Figure 1. Zero voltage resistance of the three molecules PE-n, PP-n and A-n versus their lengths.

As can be seen from Fig.1, the conjugated molecules PE-n and PP-n have much smaller resistances than equally long non-conjugated A-n molecules. It is interesting to note that the lines fitted through the data of Fig.1 converge to the resistance $R_0 = h / 2e^2 = 12.9$ kΩ which is the quantum resistance.

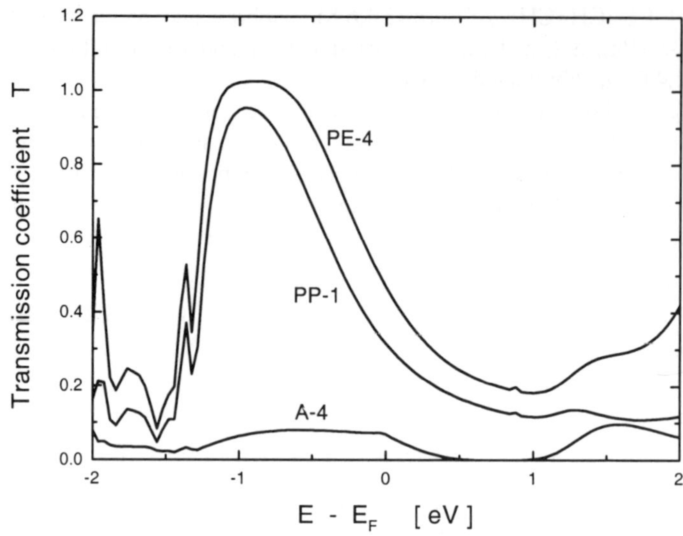

Figure 2. Transmission coefficient T(E,V=0) defined in Eq. (2) versus energy E of incident electrons for the three molecules PE-4, PP-1 and A-4 of medium length.

Fig. 2 shows the transmission coefficients $T(E,V=0)$ versus the energy E of incident electrons for the molecules PE-4, PP-1 and A-4 which have the approximate length $d_{S-S} = 0.65$ nm (see Fig. 1). The main feature of $T(E,V=0)$ in the case of the PE-4 and PP-1 molecules is a broad resonance below the Fermi level E_F centered at $E-E_F = -0.8$ eV. This broad transmission channel arises from the uppermost HOMO-n states of the H_C-Hamiltonian. In the case of the A-4 molecule Fig. 2 shows broad but weak resonances around -0.5 eV and +1.5 eV. Here the transmission mainly takes place via nonresonant states localized on the S and Au atoms. As expected, the conjugated molecules PE-4 and PP-1 have transmission coefficients that are large compared to the non-conjugated A-4 molecule.

Fig. 3 shows the transmission coefficients of the longest molecules studied $(d_{S-S} = 1.2$ nm$^{-1})$. The calculation reveals that the transmission coefficient $T(E,V=0)$ of the A-8 molecule is about two orders of magnitude smaller than that of the conjugated PE-8 and PP-2 molecules over the full energy range. In the case of the PE-8 molecule the upper HOMO-n and lower LUMO-n states of the H_C-Hamiltonian contribute to the transmission coefficient.

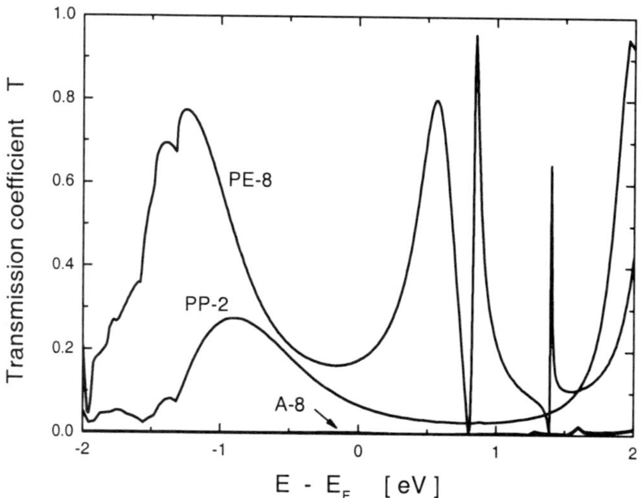

Figure 3. Transmission coefficient T(E,V=0) defined in Eq. (2) versus energy E of incident electrons for the three longest molecules PE-8, PP-2 and A-8.

Acknowledgments

We acknowledge funding from the CSIRO-ESA program.

References

1. C. Joachim, J.K. Gimzewski and A. Aviram, Nature **408**, 541 (2000).
2. M. A. Ratner, *Mater. Today, (Feb. 2002)* 20.
3. M. Di Ventra and N. D. Lang, *Phys. Rev.* **B65**, 045402 (2001).
4. Y. Xue, S. Datta, M. A. Ratner, *J. Chem. Phys.* **115**, 4292 (2001).
5. M. Brandbyge, J.-L. Mozos, P. Ordejon, J. Taylor and K. Stokbro, *Phys. Rev.* **B65**, 165401 (2002).
6. K. Stokbro, J. Taylor, M. Brandbyge, J.-L. Mozos and P. Ordejon, *Comp. Mat. Sci.* **27**, 151 (2003).
7. S. Datta, *"Electronic Transport in Mesoscopic Systems"*, Cambridge University Press: NY 1995.
8. R. E. Holmlin, R. Haag, M. L. Chabinyc, R. F. Ismagilov, A. E. Cohen, A. Terfort, M. A. Rampi and G. M. Whitesides, *J. Am. Chem. Soc.* **123**, 5075 (2001).
9. D. J. Wold and C. D. Frisbie, *J. Am. Chem. Soc.* **123**, 5549 (2001), *ibid.* **122**, 2970 (2000).

NANOBIOCERAMICS

B. Ben-Nissan [1,2]

†

[1]Department of Chemistry, Materials and Forensic Science, and

[2]Institute for Nanoscale Technology, UTS

University of Technology, Sydney, PO BOX 123 Broadway, 2007 NSW Australia,

e-mail: b.ben-nissan@uts.edu.au

Recent advances in the fundamental understanding of cell and molecular biology, tissue engineering, targeted drug delivery, wound healing and other biomedical processes, together with the development of new enabling technologies such as micro, nanoscale, and bio-inspired fabrication and surface modification methods, have the potential to drive the design and development of new biomaterials useful for medical applications at an unprecedented rate. The current focus is on the production of new nanobioceramics relevant to a broad range of applications such as implantable surface modified medical devices for better hard and soft tissue attachment, increased bioactivity, tissue regeneration and engineering, cancer treatment, drug delivery and gene therapies, treatment of bacterial and viral infections, delivery of oxygen to damaged tissues, imaging, materials for minimally invasive surgery and most promising of all nanorobotics, nanobiosensors and nanodevices for a wide range of biomedical applications.

1. Introduction

In the early 1970's, Klawitter and Hulbert proposed that bioceramics could be employed to perform singular, biologically inert roles, such as to provide parts for bone replacement. It was a pioneering approach that opened a new range of numerous biomedical applications. The realization that cells and tissues in the body perform many other vital regulatory and metabolic roles, the demands on bioceramics have changed from maintaining an essentially physical function without eliciting a host response, to providing a more integrated interaction with the host. This has been accompanied by improvements on biomimetics and self-assembly, better understanding of the bone-implant interactions and the application of new generation nanobioceramics. This better understanding and production methods lead to the design and production of new devices and

---

† This work was partially supported by the Australian Academy of Science and the Japan Society for Promotion of Science.

materials for medical devices to improve the quality of life, as well as extend its longevity.

Biomimetics refers to human-made processes, substances, devices, or systems that imitate nature or that learn from, and copy, biological systems. Biomolecular self-assembly, nano- and microfabrication, soft lithography, molecular imprinting, chemical modification and organic and bio-organic synthesis are the main current research and development areas that currently are shaping the production methods of the nanobioceramics field.

The bone mineral is composed of nanocrystals, or more accurately nano-platelets, originally described as hydroxyapatite similar to the dahllite mineral. It is now agreed that bone apatite can be better described as carbonate hydroxyapatite (CHA) and approximated by the formula: $(Ca,Mg,Na)_{10}(PO_4,CO_3)_6(OH)_2$. [1] The composition of commercial CHA is similar to bone mineral apatite. Bone pore sizes range from 1 to 100 μm in normal cortical bone and 200 to 400 μm in trabecular bone and the pores are interconnected. Porosity (macroporosity) is introduced in synthetic monolithic calcium phosphates (HAp, tri-calcium phosphate, β-TCP) by the addition of volatile compounds (e.g., naphthalene or hydrogen peroxide).

Bone-like hydroxyapatite nano powders and platelets can be synthesized with a range of production methods; however one of the most commonly used methods is from sol-gel aqueous solutions. Early published work shows that while biphasic sol-gel HAp products are easily synthesized, monophasic hydroxyapatite powders and coatings were more difficult to produce. For example, syntheses from a stoichiometric ratio of mixed salt-alkoxide sol-gel systems produce impure hydroxyapatite accompanied by a second phase of calcium oxide. Further investigations into the process showed that the time period between the mixing of precursors and heating to remove the solvent referred to as "aging time" can significantly alter the composition of the product.

It has been accepted that due to its unfavorable mechanical properties porous HAp cannot be used for load bearing applications. For this reason in orthopaedic surgery hydroxyapatite has been used as thin film coatings on metallic alloys. Of the metallic alloys used, cobalt-chromium and titanium based alloys are the preferred materials for these thin film coatings.

Thermal spraying tends to be the most commonly used coating technique but has been faced with challenge of producing a controllable resorption response in clinical situations. Other techniques that are capable of producing thin coatings include pulsed-laser deposition and sputtering which, like thermal spraying involves high-temperature processing. Other techniques capable of nanocoatings, such as electrodeposition, and sol-gel utilise lower temperatures and avoid the problems associated with the structural instability of hydroxyapatite at elevated temperatures.

The advantages of the sol-gel technique are numerous; it is nanoscale, it results in a stoichiometric, homogeneous and pure coating due to mixing on the molecular scale; it requires reduced firing temperatures due to small particle sizes with high surface areas; it has the ability to produce uniform fine-grained structures; it allows the use of different chemical routes (alkoxide or aqueous based); and it is easily applied to complex shapes with a range of coating techniques including dip, spin, and spray coating. Currently 40 nm single layer pure hydroxyapatite or carbonated apatites are easily produced. Multilayers up to 12 layers have been reported [2].

The understanding of new materials at the molecular and nanoscale level has become increasingly critical in the new era of biomedical materials for the design, and synthesis of nano-devices on the molecular scale. New technologies including molecular self-assembly as a fabrication tool will become increasingly more important in the coming decades.

Basic engineering principles for microfabrication can be learned through understanding the molecular self-assembly phenomenon. This self-assembly phenomenon is ubiquitous in nature. Coral and fascinatingly complex bone, are good examples of these biomimetic processes. Coralline apatites can be derived from the sea coral, which is composed of calcium carbonate in the form of aragonite and is a naturally occurring structure that has optimal strength and structural characteristics. The pore structure of coralline calcium phosphate produced by certain species is similar to human cancellous bone, making it a suitable material for bone graft applications. An Australian coral was converted to monophasic hydroxyapatite by using a two-stage process where the hydrothermal method was followed by a patented hydroxyapatite sol-gel nanocoating process was used [3]. This nanocoating was reported to produce a two-fold increase in the biaxial strength of the converted and nanocoated coral compared to the converted only form. Addition of bone morphogenic proteins (BMP-2), growth factors and stem cells to these and similar scaffolding materials is one of the on going research areas of the nanoceramics field.

Nanotechnologies are one of the key technologies of the 21st century. Rationally designed and synthesised materials through "bottom up" approaches can permit the tailoring of properties from the atomic to the mesoscopic and the macroscopic length scales. This ability to tailor a material's properties over broad length scales suggests that research on hybrids can significantly impact diverse fields, such as nanophotonics, separation techniques, catalysis, smart coatings, sensors, cosmetics, biomedical and ceramic and polymer composite applications.

It is believed that the synthesis and identification of properties of hybrid organic-inorganic-bio-nanocomposites, biogenic inorganic materials, bioactive inorganic and hybrid materials (encapsulation), self-assembly and templated growth processing; engineered organisation of solids at organic/inorganic interfaces will be the precursors of the future nano-sensors and nanomachines

Biomimetic or bioinspired approaches to materials and templated growth of inorganic or hybrid networks using self-assembled hybrid organic-inorganic interfaces will also help to extend this promising domain. Chemically derived hybrid materials lie at the interface of organic-inorganic and biological realms. The main advantages in the use of these hybrids result from their high versatility offering a wide range of possibilities to produce elaborate tailor-made materials in terms of chemical and physical properties. Moreover these hybrid nanocomposites present the paramount advantage of facilitating integration, miniaturisation and multi-functionalisation of the devices opening a new range of opportunities for many applications.

The current development of new novel nanoceramics are encouraging and they can be used for a broad spectrum of medical applications such as implantable medical devices, tissue engineering, drug and gene delivery, imaging agents, materials for minimally invasive surgery and biosensors. Bioceramics and specifically nanobioceramics are fundamental to the design and development of a wide variety of medical devices and implants.

References

1. R.Z. LeGeros, *Adv. Dent. Res.* **2**, 164 (1988).
2. B. Ben-Nissan, D.D. Green, G.S.K. Kannangara, C.S. Chai, and A. Milev, *J. Sol-Gel Sci. Technol.* **21**, 27 (2001).
3. B. Ben-Nissan, A. Milev, D. Green, R M Conway, G S K Kannangara, J. Russell, J. Hu, E.Gillott, and C. Trefry. International Patent, PCT -WO 02/40398 A1 (May 23, 2002).

DESIGN OF FUNCTIONAL LIPOSOMES AND THEIR APPLICATION TO DRUG AND GENE DELIVERY

KENJI KONO

Department of Applied Materials Science, Osaka Prefecture University,

1-1 Gakuen-cho, Sakai, Osaka 599-8531, Japans

Two types of functional liposomes, which exhibit temperature- or pH-sensitive properties, were designed by using synthetic polymers. Thermosensitive polymers, such as copolymers of N-isopropylacrylamide (NIPAM) and copolymers of (2-ethoxy)ethoxylethyl vinyl ether, were fixed onto the surface of liposomes. The liposomes were stabilized by highly hydrated polymer chains below its lower critical solution temperature (LCST), but destabilized by the dehydrated polymer chains above the LCST. Such an alteration of the copolymer chains on the liposome surface provided liposomes with temperature-controlled content release property and with temperature-controlled cellular affinity. Liposomes modified with succinylated poly(glycidol) (SucPG), which became fusogenic under weakly acidic conditions, were also prepared. The liposomes delivered membrane-impermeable compounds into cytoplasm of a cell by fusing with endosomal membrane. When complexed with lipoplexes, which are complexes of cationic liposomes and plasmid DNA, the SucPG-modified liposome enhanced delivery of plasmid DNA into cells by promoting fusion between the lipoplex and endosome. Thus, these temperature-sensitive and pH-sensitive liposomes may have potential usefulness for drug and gene delivery.

1. Introduction

Because liposomes can act as biocompatible and biodegradable nanocapsules, they are considered to be an ideal carrier for drugs and genes. In order to elevate their usefulness, functionalization of liposomes has been actively attempted. While several approaches have been employed, one of the most effective methods is modification of liposomes with functional polymers [1]. In this study, we designed two types of functional liposomes, namely temperature-sensitive liposomes and pH-sensitive, fusogenic liposomes by surface modification with synthetic polymers. Their preparation and functionalities are described.

2. Temperature-sensitive liposomes

There are a number of thermosensitive polymers, which exhibit LCST. These polymers are highly hydrophilic and soluble in water below the LCST. However, they become hydrophobic and water-insoluble above the LCST. To obtain liposomes with temperature-controlled functionalities, we designed liposomes modified with thermosensitive polymers, such as copolymers of NIPAM (Figure. 1) [2,3]. Egg yolk phosphatidylcholine (EYPC) liposomes modified with copolymers of NIPAM, acrylamide, and N,N-didodecylacrylamide with varying LCSTs were prepared by hydration of their mixtures and subsequent extrusion through a polycarbonate membrane with a pore size of 100 nm. The polymer-modified liposomes were very stable below the LCST of the copolymers, but were destabilized intensively above the LCST, resulting in drastic release of their contents. Because LCST of the copolymers can be controlled by composition of the copolymers, this result indicates that liposomes which release contents at a desired temperature can be produced by using the copolymer with the LCST of that temperature [4].

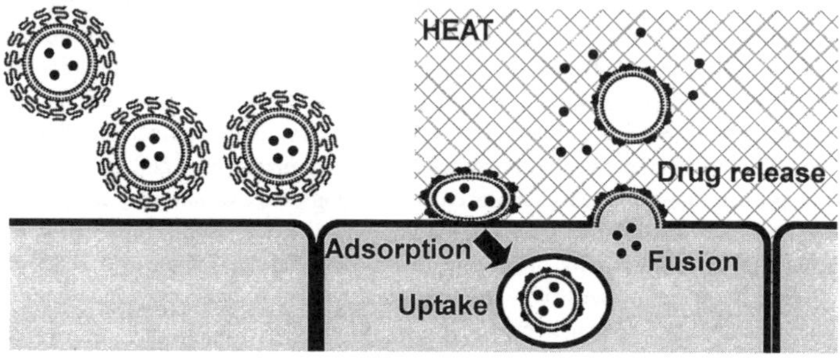

Figure 1. Functionalities of thermosensitive polymer-modified liposomes.

In order to obtain information about relationship between characteristics of thermosensitive polymers and their ability to sensitize liposomes, NIPAM copolymers, which exhibit same LCST but different transition enthalpies, were synthesized by using N-isopropylmethacrylamide (NIPMAM), N,N-dimethylacrylamide (DMAM), and N-acryloylpyrrolidine (APr) as the comonomers. These copolymers underwent a coil-to-globule transition at 40 °C, whereas their transition enthalpy increased in the order of APr-NIPAM copolymer < DMAM-NIPAM copolymer < NIPMAM-NIPAM copolymer. Liposomes having these copolymers were prepared and their contents release behavior was examined. While these copolymer-modified liposomes enhanced the release of entrapped calcein above the LCST, extent of the enhancement was quite different among these liposomes. The liposomes modified with APr-NIPAM copolymer increased the release only slightly, whereas the liposomes modified with DMAM-NIPAM copolymer showed moderate enhancement of the release. Among them, liposomes modified with NIPMAM-NIPAM copolymer exhibited most intensive release above the LCST. These results indicate that the copolymer with large transition enthalpy can effectively sensitize liposomes.

Considering the temperature-dependent conformational change of thermosensitive polymers on the surface of the liposome membrane, it is expected that surface property can be controlled by ambient temperature. In fact, liposomes having a NIPAM-APr copolymer were shown to elevate hydrophobicity of the liposome surface above the LCST and increase its affinity for cell surface [5].

Because efficiency of the transition of thermosensitive polymers depends on their chain length, their ability to sensitize liposomes would be affected by their molecular weight. Thus, block copolymers of (2-ethoxy)ethoxyethyl vinyl ether and octadecyl vinyl ether with different molecular weights were prepared by living cationic polymerization. The block of poly[(2-ethoxy)ethoxyethyl vinyl ether] acts as a temperature-sensitive moiety with the transition temperature of ca. 35 °C, whereas the block of poly(octadecyl vinyl ether) acts as an anchoring moiety. The copolymer having a higher molecular weight covered the liposome surface more efficiently below the LCST and underwent the conformational transition in a narrower temperature region, resulting in more effective control of surface properties of liposomes.

Temperature-sensitivity of thermosensitive polymer-modified liposomes has been shown to be affected by various factors, such as lipid composition [6] and attachment of poly(ethylene glycol) grafts to liposome surface [7], in addition to structure and characteristics of the polymers used for the modification. By optimizing these factors, highly sensitive liposomes can be produced [7].

3. *pH-Sensitive liposomes*

pH-Sensitive liposomes are liposomes that become fusogenic under weakly acidic conditions. When taken up by a cell through endocytosis, pH-sensitive liposomes can fuse with the membrane of endosome, which contains a mildly acidic environment, and introduce contents into cytosol. Therefore, pH-sensitive liposomes are considered to be useful for cytoplasmic delivery of membrane-impermeable molecules.

We prepared pH-sensitive liposomes by surface modification of liposomes with a pH-sensitive, fusogenic polymer, SucPG (Figure 2) [8]. Carboxylate anions of the polymer chain suppress interaction of the polymer chain with liposomal membrane at neutral pH. However, their protonation enhances interaction of the polymer chain with the liposomal membrane, inducing destabilization of the liposome. Fusion ability of SucPG-modified EYPC liposomes rose significantly below pH 6.0, which is roughly same to the pH of endosomal environment. When CV1 cells were treated with SucPG-modified liposomes encapsulating calcein, we observed that the cells displayed diffuse

146

fluorescence. This is because calcein molecule was introduced into cytoplasm via fusion between the SucPG-modified liposome and endosomal membrane [9].

For gene therapy, it is desired to develop vectors, which introduce therapeutic genes into target cells and induce their efficient transfection. So far, cationic liposomes and cationic polymers have been used for this purpose. These cationic systems can interact with DNA through electrostatic interactions and form complexes, which strongly interact with cells and introduce DNA into them. However, these systems can not transfect cells as efficiently as viral vectors do. One of the problems for these cationic vector systems to be solved is that after internalization, a large fraction of the DNA-vector complexes are degraded in lysosome. Therefore, escape of gene from the degradation in the lysosomal compartment is considered to be the key for the efficient transfection.

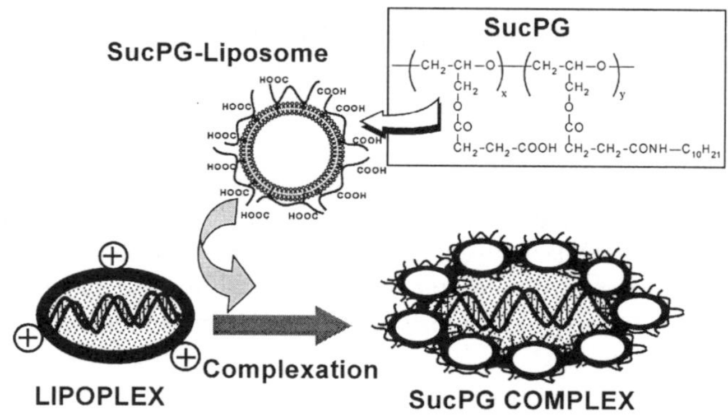

Figure 2. Preparation of SucPG complex.

To obtain a synthetic vector, which achieves efficient gene transfection, we designed a complex of SucPG-modified liposome and lipoplex, which is termed SucPG complex (Figure 2), because SucPG-modified liposome fixed onto the surface of the lipoplex can enhance transfer of genes by promoting fusion between the lipoplex and endosomal membrane. The SucPG complex was prepared by mixing SucPG-modified liposomes bearing transferrin and DC-chol-based lipoplex. The SucPG complex effectively transfected various cancer cells, such as HeLa and K562 cells [10]. This transfection was strongly suppressed in the presence of free transferrin, indicating that the SucPG complex was taken up by the cells through the specific binding between transferrin and its receptor. In addition, the SucPG complex was shown to have low toxicity and high serum resistance, compared with conventional cationic liposomes, probably due to its negatively charged surface.

It is expected that the transfection activity of SucPG complex can be improved by adjusting its composition, charge ratio, and size. Also, selection of cationic lipid component may improve its activity. Thus, after optimization of the structure, the SucPG complex may be used as an efficient nonviral vector.

References

1. H. Ringsdorf, B. Schlarb, J. Venzmer, *Angew. Chem. Int. Ed. Engl.*, **27**, 113 (1988).
2. K. Kono, H. Hayashi, T. Takagishi, *J. Control. Release*, **30**, 69 (1994).
3. K. Kono, *Adv. Drug Delivery Rev.*, 53, 307 (2001).
4. H. Hayashi, K. Kono, T. Takagishi, *Bioconjugate Chem.* **10**, 412 (1999).
5. K. Kono, R. Nakai, K. Morimoto, T. Takagishi, *FEBS Lett.,* **456**, 306 (1999).
6. K. Kono, A. Henmi, H. Yamashita, T. Takagishi, *J. Control. Release*, **59**, 63 (1999).
7. K. Kono, K. Yoshino, T. Takagishi, *J. Control. Release*, **80**, 321 (2002).
8. K. Kono, K. Zenitani, T. Takagishi, *Biochim. Biophys. Acta*, 1193, 1 (1994).
9. K. Kono, T. Igawa, T. Takagishi, *Biochim. Biophys. Acta*, 1325, 143 (1997).
10. K. Kono, Y. Torikoshi, M. Mitsutomi, T. Itoh, N. Emi, T. Yanagie, T. Takagishi , *Gene Ther.,* **8**, 5-9 (2001).

TARGETING OF ^{10}B COMPOUNDS TO SOLID TUMORS BY TRANSFERRIN-PEG LIPOSOMES, FOR BORON NEUTRON-CAPTURE THERAPY (BNCT)

KAZUO MARUYAMA[1], S. KASAOKA[1], T. TAKIZAWA[1], N. UTOGUCHI[1], , Y. SAKURAI[2], K. ONO[2], H. KOBAYASHI[3], A. SHINOHARA[4], H. YANAGIE[5]

[1]*Dept. of Biopharmaceutics, Teikyo Uinv. Kanagawa;* [2]*Research Reactor Institute, Kyoto Univ. Kyoto;* [3]*Institute for Atomic Reactor, Rikkyo Univ. Kanagawa; and* [4]*Dept of Epidemiology, Juntendo University School of Medicine, Tokyo, Japan;* [5]*RCAST, Univ. of Tokyo, Tokyo;*

The successful treatment of cancer by boron neutron-capture therapy (BNCT) requires the selective delivery of relatively high concentration of ^{10}B compounds to malignant tumor tissue. This study focuses on a new tumor-targeting drug delivery system for BNCT that uses small (-200 nm in diameter), unilamellar mercaptoundeca hydrododecaborate (BSH)-encapsulating, transferrin (TF)-conjugated polyethylene-glycol liposomes (TF-PEG liposomes). These liposomes were prepared from DSPC, CH, DSPE-PEG and DSPE-PEG-COOH (2:1:0.11:0.021, molar ratio), and each liposome bears approximately 20 TF molecules conjugated via the carboxyl residue of DSPE-PEG-COOH. We examined the antitumor activities of liposomal BSH in combination with thermal neutron irradiation in Colon 26 tumor-bearing mice. When TF-PEG liposomes were injected at a dose of 35 mg ^{10}B/kg, we observed a prolonged residence time in the circulation and low uptake by the reticuloendothelial system (RES) in Colon 26 tumor-bearing mice, resulting in enhanced accumulation of ^{10}B into the solid tumor tissue (e.g., 35.5 μg/g). TF-PEG liposomes maintained a high ^{10}B level in the tumor, with concentrations over 30 μg/g for at least 72 hours after injection. This high retention of ^{10}B in tumor tissue indicates that binding and concomitant cellular uptake of the extravasated TF-PEG liposomes occurs by TF receptor and receptor-mediated endocytosis, respectively. On the other hand, the plasma level of ^{10}B decreased, resulting in a tumor/plasma ratio of 6.0 at 72 hours after injection. Therefore, 72 hours after injection of TF-PEG liposomes was selected as the time point of BNCT treatment. Thus, intravenous injection of TF-PEG liposomes can increase the tumor retention of ^{10}B atoms, which were introduced by receptor-mediated endocytosis of liposomes after binding, causing tumor growth suppression *in vivo* upon thermal neutron irradiation. These results suggest that BSH-encapsulating TF-PEG liposomes may be useful as a new intracellular targeting carrier in BNCT therapy for cancer.

1. Introduction

Boron neutron capture therapy (BNCT) is aimed at inhibiting the growth of various cancers [1]. The stable isotope of boron, ^{10}B, interacts with low energy (thermal) neutrons to produce highly energetic, short-range disintegration products. These particles (α and ^{7}Li) destroy cells within about 10 μm from the site of the capture reaction. It is theoretically possible to kill tumor cells without affecting adjacent healthy cells, if ^{10}B atoms can be selectively accumulated in the interstitial space of tumor tissue and/or intracellular space of tumor cell. Thus, successful treatment of cancer by BNCT requires the selective delivery of relatively large amounts of ^{10}B compounds to malignant cells. The estimated boron concentration required for effective therapy is in the range of 20-30 μg ^{10}B per g of tissue [2]. At the same time, the boron concentration in the surrounding normal tissue should be kept low to minimize damage to the normal tissue.

We have recently demonstrated that transferrin-coupling pendant-type PEG liposomes (TF-PEG liposomes) were extravasated effectively into solid tumor tissue in Colon 26 tumor-bearing mice, and internalized into tumor cells [3]. TF-PEG liposomes showed a prolonged residence time in the circulation and low RES uptake in tumor-bearing mice, resulting in enhanced extravasation of the liposomes into the solid tumor tissue. This phenomenon has been characterized and termed the tumor-selective enhanced permeability and retention (EPR) effect of macromolecules and lipidic particles including liposomes [4, 5]. This predominant tumor accumulation was due to the unique vascular characteristics of the tumor tissue such as hypervasculature and enhanced vascular permeability, and the lack of a lymphatic recovery system in the solid tumor [6, 7]. Once at the tumor site, TF-PEG liposomes were internalized into tumor cells by receptor-mediated endocytosis. TF-PEG liposomes were taken up into endosome-like intracellular vesicles. It is well known that the TF receptor concentration on tumor cells is much higher than that on normal cells (for review, see ref [8]). Transferrin receptor-mediated endocytosis is a normal physiological process by which transferrin delivers iron to the cells [9, 10]. Therefore, the clearance of TF-PEG-liposomes from tumor tissue is so impaired that they remain in the tumor interstitium for a long time [3].

In this study, we first examined the potential of liposomes for selective delivery of therapeutic quantities of ^{10}B to tumors. TF-PEG liposomes and PEG liposomes encapsulating BSH were prepared and their tissue distributions in Colon 26 tumor-bearing mice after i.v. injection were compared with those of bare liposomes and free BSH. On the basis of the findings in biodistribution studies, we selected a suitable dosage of BSH and a time point for thermal neutron irradiation for BNCT. We then examined the antitumor activities of liposomal BSH in combination with thermal neutron irradiation under the selected conditions, in Colon 26 tumor-bearing mice.

2. Material and methods

2.1. *Lipids and chemicals*

Sodium mercaptoundecahydrododecaborate, $Na_2B_{12}H_{11}SH$ (BSH, M.W.=208.9, ^{10}B enriched >99 %) was purchased from Ryscor Science Inc. (Raleigh, NC). Human iron-saturated TF was purchased from Sigma Chem. (St. Louis, MI). DSPC (COATSOME MC–8080), DSPE (COATSOME ME–8080), monomethoxy polyethyleneglycol succinimidyl succinate (PEG-OSu), and polyethyleneglycol bis(succinimidyl succinate) (PEG-2OSu) were kindly donated by Nippon Oil and Fats Co. (Tokyo, Japan). The values of the number-average molecular weight of PEG-OSu and PEG-2OSu were 2219 and 3230.

2.2. *Preparation of TF-pendant-type PEG liposomes encapsulating BSH*

Bare liposomes and PEG liposomes were prepared from DSPC and CH (2:1, molar ratio) and DSPC, CH, DSPE-PEG and DSPE-PEG-COOH (2:1:0.11:0.021, molar ratio), respectively. Small unilamellar vesicles (SUV) of these two liposomes were prepared according to the reverse-phase evaporation (REV) method. TF-pendant-type PEG liposomes (TF-PEG liposomes) were prepared by coupling of TF to PEG liposomes as described previously [3].

2.3. *Biodistribution in tumor-bearing mice*

Tumor-bearing mice were prepared by inoculating s.c. a suspension (5×10^6 cells) of Colon 26 cells directly into the back of male BALB/c mice (7 weeks old, weighing 20-23 g, Nihon SLC, Tokyo). Two hundred to 300 μl of BSH solution, BSH bare liposomes, BSH-PEG liposomes or BSH-TF-PEG liposomes was injected into tumor-bearing mice (3-5 per group) via the tail vein at a selected dose of ^{10}B.

2.4. *BNCT for tumor-bearing mice*

^{10}B-solution, ^{10}B-containing bare liposomes, ^{10}B-PEG liposomes or ^{10}B-TF-PEG liposomes were injected into tumor-bearing mice (7-10 per group) via the tail vein at a dose of 35, 20 or 5 mg ^{10}B/kg. At 72 hours after administration, the mice were anesthetized with sodium pentobarbital solution and placed in an acrylic mouse holder (Iuchi, Tokyo). The mice were irradiated in the TRIGA-II atomic reactor for 37 min at a rate of 2×10^{12} neutrons/cm^2. The antitumor effect of BNCT was evaluated on the basis of the changes in tumor volume and survival of the mice. The mortality was monitored daily and body weight and tumor volume was measured at intervals of a few days. For determining the tumor volume, two perpendicular diameters of the tumor were measured with a slide caliper, and calculation was carried out using the formula $0.5 (A \times B^2)$, where A and B are the longest and shortest dimensions of the tumor in millimeters, respectively.

3. Results and Discussion

3.1. *Characterization of liposomal formulations containing BSH*

Homogeneous conventional bare liposomes and PEG liposomes composed of DSPC, CH (2:1, molar ratio) and DSPC, CH, DSPE-PEG and DSPE-PEG-COOH (2:1:0.11:0.021, molar ratio), with an average diameter of 107-117 nm, were prepared by an REV method followed by extrusion. The ^{10}B content per μmol lipid was 26–30 μg. TF-PEG liposomes containing BSH, bearing approximately 20 TF molecules per liposome, were prepared by coupling of TF to the extremities of the PEG chains of PEG liposomes. The leakage of BSH from PEG liposomes was very small when TF was coupled at low levels, such as less than 1 %. The change of size of liposomes and the release of BSH from the three different BSH liposomal formulations were small during storage for 50 days in refrigerator. Because the osmotic pressure of 125 mM BSH solution is 317 mOsm, approximately equal to that of serum or 9 % sucrose solution, the liposomes were extremely stable in 9 % sucrose solution when stored at low temperature.

3.2. *Binding of TF-PEG liposomes and uptake of BSH by cells in vitro*

The specificity of the TF-receptor mediated binding of TF-PEG liposomes (an average of 20 TF molecules per liposome) encapsulating BSH was studied. The data was presented as an associated ^{10}B concentration (Fig. 1). The data at 4°C in Fig. 1 indicated that TF-PEG liposomes readily bound with Colon 26 cells in the medium containing 10% FCS. The ^{10}B levels in TF-PEG liposomes were greater than that in nontargeted PEG liposomes and bare liposomes. The presence of 50 μg of free TF decreased significantly ^{10}B level in TF-PEG liposomes. These results revealed that the binding of TF-PEG liposomes to Colon 26 cells is indeed receptor specific, and free PEG (not linked to antibody) in liposomes does not interfere

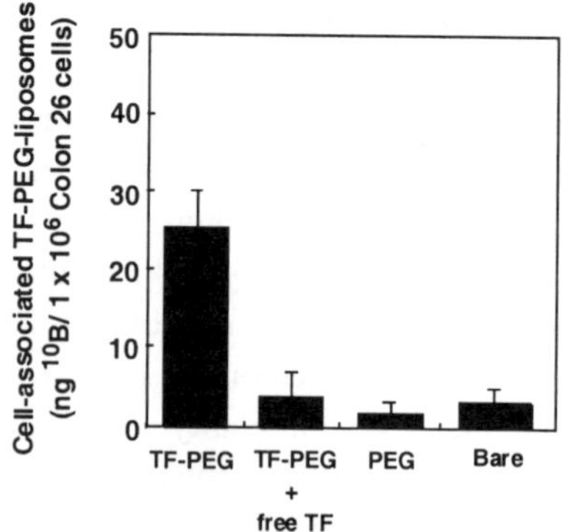

Fig. 1 Receptor specific binding and internalization of TF-PEG liposome to Colon 26 cells.

sterically with TF-receptor binding, since TF is covalently linked to the distal terminal of PEG chains on the external surface of PEG liposomes.

3.3. *Biodistribution of BSH loaded in liposomes in tumor-bearing mice*

Liposomes encapsulating BSH, with an average diameter of 105-125 nm, were injected intravenously into Colon 26 tumor-bearing mice, and the distribution and tumor accumulation of ^{10}B were evaluated. Fig. 2A shows the time course of blood clearance of ^{10}B after administration of TF-PEG liposomes, PEG liposomes, bare liposomes and BSH solution at a dose of 35 mg ^{10}B/kg. In the case of i.v. administration of BSH solution, disappearance of ^{10}B from the blood circulation was very rapid due to high renal clearance. At 1 hour after injection, in spite of the high dose of 35 mg/kg, a plasma concentration of only 20 μg/ml was observed. In contrast, the plasma levels of ^{10}B encapsulated in liposomes remained high for a long time. In particular, the plasma concentrations of BSH loaded in TF-PEG liposomes and PEG liposomes were much higher than that of ^{10}B loaded in conventional liposomes (bare liposomes) for 48 hours after injection. These results confirmed the effectiveness of the PEG layer in prolonging the systemic circulation time of liposomes after i.v. injection. ^{10}B loaded in TF-PEG liposomes, bearing approximately 20 TF molecules per liposome, showed prolonged residence in the circulation and low liver uptake, as did ^{10}B loaded in PEG liposomes without TF. Thus, the conjugation of TF to the PEG terminal did not alter the RES uptake of PEG liposomes.

Fig. 2B shows the time course of ^{10}B accumulation in Colon 26 solid tumor tissue after injection of TF-PEG liposomes, PEG liposomes, bare liposomes and BSH solution at a dose of 35 mg ^{10}B/kg. The ^{10}B concentration in the tumor

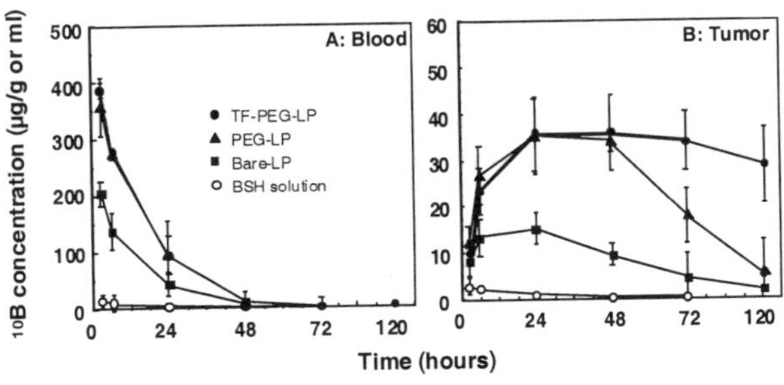

Fig. 2 Time courses of blood residence (Panel A) and tumor accumulation (Panel B) of ^{10}B delivered by each kind of liposomes encapsulating BSH and by BSH solution in Colon 26 tumor-bearing mice. Two hundred to 300 μl of BSH-TF-PEG liposomes, BSH-PEG liposomes, BSH bare liposomes or BSH solution was injected into tumor-bearing mice *via* the tail vein at a dose of 35 mg/kg.

tissue at 1 hour after injection of free BSH was 1.0 μg/g, and did not increase thereafter. Administration of liposomal BSH produced increased ^{10}B accumulation into solid tumor tissue compared with free BSH. The highest ^{10}B levels in the tumor were obtained with PEG liposomes and TF-PEG liposomes at 24 hours post injection. The ^{10}B concentration in tumor tissue was 35.5 μg/g or 35.2 μg/g in TF-PEG liposomes or PEG liposomes, respectively, and was 2.2- or 2.1-fold greater than that of bare liposomes, respectively.

The most interesting feature of these results is that when BSH loaded in TF-PEG liposomes was injected, a high ^{10}B level (over 30 μg/g) was maintained in the tumor for a longer period, as observed at 48-72 hours after injection. This high retention of ^{10}B in tumor tissue indicates cellular uptake of TF-PEG liposomes by TF receptor-mediated endocytosis. Another important point was that the plasma level of ^{10}B was very low at 48-120 hours after injection of TF-PEG liposomes. The long retention caused by TF allowed sufficient time for the plasma ^{10}B concentration to reach a low level, resulting in a tumor/plasma ratio of 6.0 at 72 hours after injection. This represents a potentially useful accumulation of ^{10}B in the tumor in conjunction with an extremely low plasma ^{10}B concentration. TF-PEG liposomes provide a pathway for selective introduction of ^{10}B inside tumor cells and offer an advantageous tumor/plasma ^{10}B ratio for BNCT. These characteristics are therapeutically favorable. Thus, 72 hours or later after injection of TF-PEG liposomes was selected as a suitable time point for BNCT treatment.

3.4. *Therapeutic effect of BNCT in tumor-bearing mice*

Since the plasma ^{10}B concentration was very low at 72 hours after injection of TF-PEG liposomes, while the ^{10}B concentration in tumor tissue remained sufficiently high, this time point was selected for BNCT treatment. The anti-tumor activity of liposomally delivered BSH in conjunction with thermal neutron irradiation was evaluated in Colon 26 tumor-bearing mice. When transplanted Colon 26 tumor cells were growing logarithmically, mice were given saline as a control, free BSH or liposomal BSH at a dose of 20mg ^{10}B/kg. Tumor growth

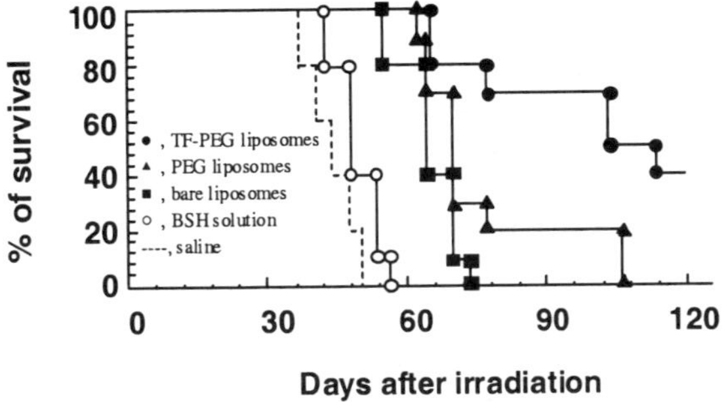

Fig. 3 Survival curve of mice bearing Colon 26 solid tumors after i.v. injection of 20 mg ^{10}B/kg of liposomal BSH or BSH solution and, at 72 hours after administration, thermal neutron irradiation with 2 x 10^{12} thermal neutrons/cm^2 for 37 min.

curves and survival curves are presented in Fig. 3. Administration of BSH encapsulated in TF-PEG liposomes resulted in superior long-term survival compared with the case of PEG liposomes. These results suggest that extravasated TF-PEG liposomes bound effectively to the tumor cell membrane and delivered BSH into the cytoplasm via receptor-mediated endocytosis. PEG liposomes were less effective than TF-PEG-liposomes, although their therapeutic index is still superior to that of free BSH.

4. Conclusion

The successful treatment of cancer by BNCT requires the selective delivery of relatively high concentration of ^{10}B compounds to malignant tumor tissue. The results presented here indicate that TF-PEG liposomes encapsulating BSH satisfied these requirements and should be very useful for BNCT of cancer. Intravenous administration of TF-PEG-liposomes with an average diameter of 100-200 nm to the tumor-bearing mice resulted in increased ^{10}B accumulation and retention time in solid tumor tissue. These phenomena are referred to as the EPR effect in solid tumor tissues and the cellular uptake of TF-PEG liposomes by TF receptor-mediated endocytosis. The extravasated TF-PEG liposomes bound effectively to the tumor cell membrane and delivered BSH into the cytoplasm via receptor-mediated endocytosis. These represent a potentially useful accumulation of ^{10}B in the tumor in conjunction with an extremely low plasma ^{10}B concentration. Thus, TF-PEG liposomes could be useful as a new intracellular targeting carrier in BNCT therapy for cancer.

5. Acknowledgements

A part of this work was supported by Grants-in-Aid for Scientific Research (No. 13470190 to Kazuo Maruyama, No. 11691202 and 11557092 to Hironobu Yanagie) from the Ministry of Education, Culture, Sport, Science and Technology, Japan and a Grant-in-Aid for Cancer Research (No. 12-1 to Kazuo Maruyama) from the Ministry of Health, Labour and Welfare, Japan.

References

1   M.F. Hawthorne, The role of chemistry in the development of boron neutron capture therapy of cancer, Angew Chem. Int. Ed. Engl. 32 (1993) 950-984.

2   R.F.Barth, A.H. Soloway, R.G. Fairchild, R.M. Brugger, Boron neutron capture therapy for cancer, Cancer 70 (1992) 2995-3007.

3   O. Ishida, K. Maruyama, H. Tanahashi, M. Iwatsuru, et al., Liposomes bearing polyethyleneglycol coupled transferrin with intracellular targeting property to the solid tumors in vivo, Pharm. Res. 18 (2001) 1042-1048.

4   Y. Matsumura, H. Maeda, A new concept for macromolecular therapeutics in cancer chemotherapy: Mechanism of tumoritropic accumulation of proteins and the antitumor agent smancs, Cancer Res. 46 (1986) 6387-6392.

5   O. Ishida, K, Maruyama, K, Sasaki, M. Iwatsuru. Size-dependent extravasation and interstitial localization of polyethyleneglycol liposomes in solid tumor-bearing mice, Int. J. Pharm. 190 (1999) 49-56.

6   R.K. Jain, L.E. Gerlowski, Extravascular transport in normal and tumor tissue, Crit. Rev. Oncol. Hematol. 5 (1986) 115-170.

7   H.F. Dvorak, J. Nagy, J. Dvorak, Identification and characterization of the blood vessels of solid tumors that are leaky to circulating macromolecules, Am. J. Pathol. 133 (1988) 95-109.

8   E. Wagner, D. Curiel, M. Cotton, Delivery of drugs, proteins and genes into cells using transferrin as a ligand for receptor-mediated endocytosis, Advanced Drug Delivery Reviews 14 (1994) 113-135.

9   H.A. Huebers, C.A. Finch, The physiology of transferrin and transferrin receptors, Physiol. Rev. 67 (1987) 520-582.

10 P. Aisen, The transferrin receptor and the release of iron from transferrin, Adv. Exp. Med. Biol. 365 (1994) 31-40.

ELECTRONIC PROPERTIES OF MOLECULES MEASURED WITH SUBMICRON-GAP ELECTRODES

W. MIZUTANI,* Y. NAITOH*, K.TSUKAGOSHI [†]

Nanotechnology Research Institute, National Institute of Advanced Industrial Science and Technology (AIST), AIST Tsukuba Central 4, 1-1-1 Higashi, Tsukuba, Ibaraki 305-8562, Japan
E-mail: W.Mizutani@aist.go.jp

Electronic properties of single molecules have been studied using newly developed electrodes with molecular-scale gaps. We are measuring the current-voltage $(I-V)$ characteristics of various molecules with sulfur moieties at the both ends. The molecular wires were synthesized to bridge the gaps of the electrode, but the direct observation of the molecules in the gap was difficult. We used synthesized DNAs with a thiol binding site, and formed direct bridges between the gap. In addition, we used the same condition to make indirect bridges with gold nanoparticles inside to visualize the DNA binding sites. We compared the $I-V$ curves of the two types of the bridges, and discuss the mechanism of the conduction.

1. Introduction

Electronic properties of molecules have been studied with electrodes sandwiching bulk crystals or thin films. Scanning probe techniques using conductive probes were introduced to estimate the local conductivity of molecular films, i.e. self-assembled monolayers, on metals with nanoscale resolution. For example, electric conductance and current-voltage $(I - V)$ characteristics of single (or several) molecules embedded in insulating matrix molecular films such as alkane thiols [1,2]. For utilizing functional molecules as a component of electronic devices, however, electrodes separated by molecular-scale gaps are required.

We examined field-induced conductance change of thin organic films

---

*Research Consortium for Synthetic Nano-Function Materials Project, AIST, 1-1-1 Umezono, Tsukuba, Ibaraki 305-8568, Japan

[†]Japan Science and Technology corporation, 4-1-8, Honcho, Kawaguchi, Saitama 332-0012, Japan. Work partially supported by NEDO under the Nanotechnology Materials Program and also supported by MEXT under the Nanotechnology Partnership Program.

using the trench-type electrodes, where the gate field is applied from the sidewall of the trench. We fabricated the electrodes, and observed weak field effects of a conductive self-assembled film and oligothiophene thin films [3].

Recently, we developed a high yield production scheme for sub-100 nm gap electrodes without using electron beam lithography [4]. We confirmed that the electrodes can be successfully used to measure the conductivity of molecular wires.

Among the studies of molecular wires, electronic conduction along DNA is a controversial issue. For DNA wires longer than 1 μm, many reports concluded that DNA molecules are insulating [5,6,7], but it is still unclear for DNA wires shorter than 100 nm. We used synthesized DNAs with a thiol binding site, and formed direct bridges between the gap. In addition, we fabricated DNA-gold nanoparticle system to identify DNAs in the nano-gap. Observed gold nanoparticles indicate the DNA binding.

We compared the $I - V$ curves of the direct DNA bridge and bridge with nanoparticles, and discuss the conduction mechanism.

2. Experimental

The guanine (G) base is known to have the highest molecular orbital (the lowest ionization potential (IP)) of four bases [8]. From experimental and theoretical studies, the contribution of G base sequences like GG and GGG to the charge transfer (the hole conduction) were investigated. It should be noted that DNA with too many G bases is difficult to synthesize and hybridize, because G–cytosine (C) interaction is so strong that they often make aggregated structures. Therefore, we designed DNA oligomers which hybridize uniquely with their pairs and contain as many G bases as possible. For a unique hybridization, we avoided repeating the same sequence in the part of DNA (Fig. 1(a)). The thiol moiety was attached at the 5' end for binding to gold surfaces. All DNAs were dissolved in Tris-EDTA buffer (pH 8.0).

We used electrodes fabricated on SiO_2 by the shadow evaporation of gold [4], whose gap widths are controlled from 5–20 nm by the evaporation angle. We checked electric conductance of the bare electrodes before use, and confirmed no detectable currents without DNA adsorption. We tried to fabricate two types of DNA bridge structures using self-assembled technique, namely, by incubating the nano-gap electrodes in a solution containing DNA thiolates. One is the direct bridging (Fig. 1(b)), and the other is via gold nanoparticles(Fig. 1(c)). For the direct bridging, A and T-type

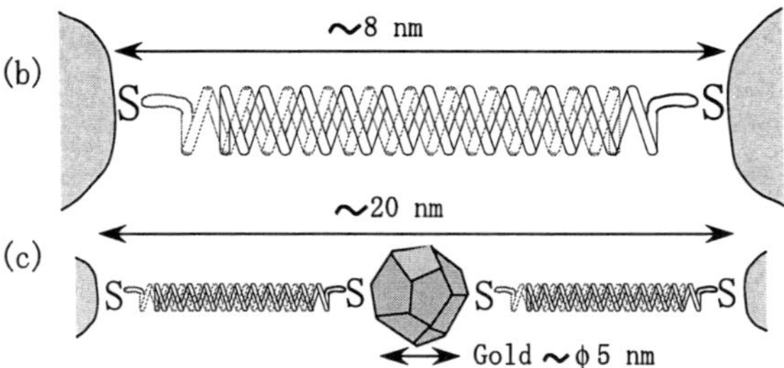

Figure 1. (a) Thiolated single stranded DNA sequences (25 bases, about 7.5 nm). A-type and its complementary T-type were used. (b) Nano-gap electrodes are directly connected by hybridized A- and T-type DNAs. (c) Electrodes connected by DNA-gold nanoparticle(5 nm in diameter).

DNA single-stranded (ss) oligomers were hybridized in the solution, and then applied to the electrodes. Hybridization was performed in the hybridization chamber at 80°C for 10 min, followed by standing for 1 h at room temperature. We could also adsorb one type of ss-DNA to the electrodes first, and hybridize the complementary ss-DNA afterwards. After the DNA adsorption, we added a 1 mM aqueous solution of OH-terminated short alkanethiol (OH(CH$_2$)$_6$SH) to fill the rest of the electrode surfaces, which is expected to prohibit DNA from laying down, and provide some space for DNA wires to move for reaction. After the incubation, the substrates were rinsed with pure water and dried under N$_2$ gas flow.

In order to confirm DNA adsorption on the electrodes and in the gap, we used gold nanoparticles as a marker. Gold nanoparticles and the nano-gap electrodes were separately coated with T-type and A-type DNA thiolates (Fig. 1(a)), respectively. The electrode was treated with the alkanethiol solution, then the solution containing the nanoparticle-DNA was applied on the nano-gap electrodes, and annealed for hybridization. We confirmed that the gold nanoparticles with thiolated DNAs adsorbed on the electrodes covered with the complementary DNA under the same adsorption conditions used in this work(Fig. 2).

The $I-V$ characteristics of the DNA bridges were then measured with a prober system at room temperature and low temperatures of 130–170 K in vacuum. Measured electronic conduction of DNA bridges was not stable at room temperature, suggesting thermal fluctuation of the DNA binding

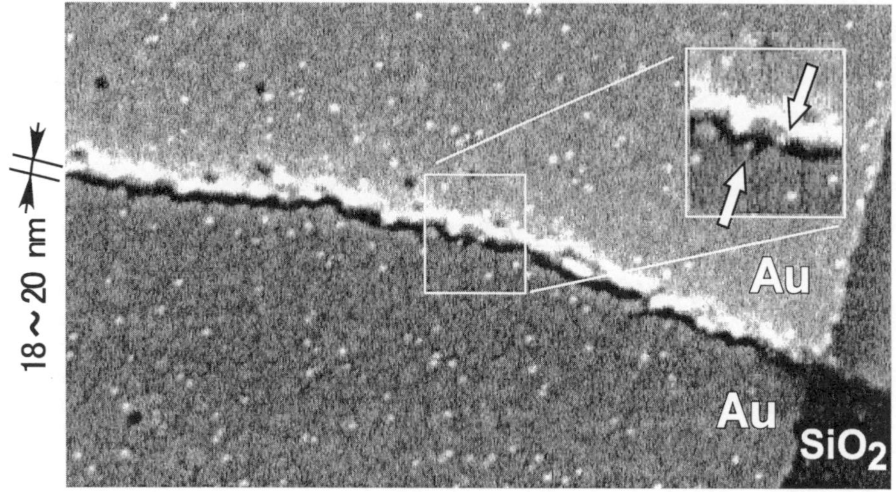

Figure 2. Scanning electron microscopy image of DNA-nanoparticle adsorption on a nano-gap electrode. White dots represent gold nanoparticles of 5 nm in diameter, and narrow black line is the nano-gap. Inset shows magnified image of the white rectangle. Arrows indicate nanoparticles in the gap.

sites and/or ionic conduction through the adsorbed water remaining even after the drying and evacuation process. The $I - V$ curves measured at low temperature were more reproducible than those measured at room temperature, although abrupt changes as shown in Fig. 3(a) were sometimes observed.

3. Results and Discussion

3.1. *Electronic states of DNA bases*

We investigated the electronic states of DNA to understand the measured $I - V$ characteristic. Voityuk *et al.* [9] calculated IP of DNA bases. For example, IP of G was estimated to be 8.098 eV (theoretical) and 8.24 eV (experimental). They reported further that the IPs decrease by 0.3 eV when base pairs (G–C, adenine (A)–thymine (T)) were formed. Lewis *et al.* estimated that the IPs of G, GG, and GGG are 7.51, 7.28, and 7.07 eV, respectively[10]. Yoshioka *et al.* estimated that the IPs of TGGG and CGGG are 6.45 and 6.39 eV, respectively [8]. According to these estimation, we assumed that GC rich DNA double stranded wires have the IP of 7.5±1 eV. If we neglect the complex problem of the DNA-electrode interface, it is reasonable for DNA wires to show the current onset at around 2 V, since

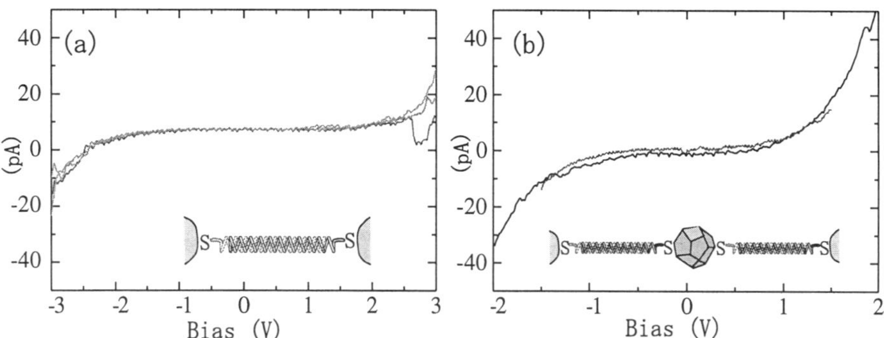

Figure 3. (a) $I - V$ characteristics of nano-gap electrodes directly connected by A- and T-type DNAs measured in vacuum at 170 K. The measurements were repeated several times with the same sample to confirm reproducibility (3 scans were plotted). (b) $I - V$ characteristics of the electrodes connected by DNA-gold nanoparticle(5 nm in diameter). Measurements were repeated several times, and 2 scans are shown. The onset voltage is lower than the direct DNA bridges (a)

the work-function of gold electrodes is about 5.1 eV.

More accurate simulation was made by Hjort and Stafström using poly(G)-poly(C) DNA wire with the platinum contacts (work-function of 5.6 eV) [11]. They assumed the resonant tunneling process via G orbital for the DNA conduction, and obtained an $I - V$ curve with the onset voltage of about 1.8 eV, which agrees well with the experiments by Porath et al.[12] and by us.

3.2. *DNA-nanoparticle system*

The onset voltage of the $I - V$ measured for DNA-nanoparticle junctions is smaller than the direct bridges. If the same conduction mechanism worked for the both structures, the former should require the double voltages of the latter because of the series connection. We have to consider other mechanisms contributing to the nanoparticle system. One possibility is the single electron tunneling. We calculated the $I - V$ using the equations proposed by Amman et al. [13] and Hanna and Tinkham [14]with parameters adjusted to reproduce the experimental results, and plotted a theoretical curve to compare with the experimental curve as shown in Fig. 4. The capacitance of $10^{-20}-10^{-19}$ F is a reasonable order for the nanoparticle of 2.5 nm in radius locating about 8 nm apart from the electrodes.

The absolute resistance of the single DNA wire is difficult to estimate, because the total number of the wires connecting the electrodes is not

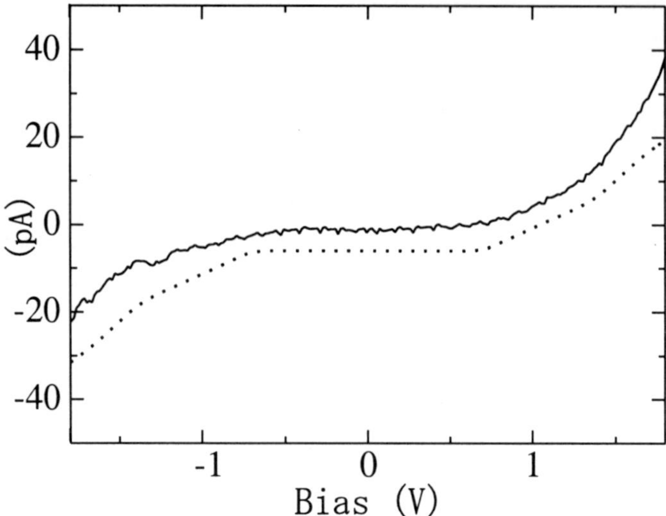

Figure 4. Experimental $I - V$ curve of DNA-nanoparticle connection (the same curve as the one in Fig. 3(b)), and theoretical $I - V$ curve (dotted line) calculated using the parameters (C_1=1.4×10^{-19}F, C_2=7×10^{-20}F, R_1=10$^{11}\Omega$, and R_2=5×10$^{11}\Omega$, T=160 K. Plotted with vertical magnification(×14) and downward offset for clarity). Parameters were not optimized to fit the experimental curve.

known. If DNA thiolates adsorbed on the surface of gold nanoparticle (ϕ5 nm) as densely as alkanethiols on the gold (111) surface, the total number of adsorbed molecules is about 360. We can only guess that the number of the DNA wires contributing to the conduction per nanoparticle is less than a hundred, probably 10-50. As can be seen in Fig. 2, we observed a couple of nanoparticles in the gap. Effects of parallel connection are under investigation. We speculate that the DNA-nanoparticle system could enhance the conductivity, e.g., by connecting the gaps at the parts where the width is shorter than 15 nm, or because the nanoparticle with DNA thiolates acted as a moving electrode allowing more effective hybridization than bridging the fixed electrodes.

4. Summary

The electric properties of DNA molecules were measured using fixed electrodes with molecular-scale (nanometer) gaps. We designed and synthesized G-C rich DNA oligomers with sulfur binding sites. In the case of electrodes directly connected by thiolated DNAs, current-voltage characteristics with an onset at about 2 V were observed. When the electrode gap was connected by DNA-gold nanoparticle-DNA, the current onset was decreased to around 1 V. We assume the single-electron tunneling phenomenon to account for this small onset.

Acknowledgement

We are grateful to H. Azehara and T. Ishida for the valuable discussion. We also thank M. Machida, S. Kunihiro, K. Tanaka and M. Shionoya for their kind supports for DNA characterization.

References

1. X. D. Cui, A. Primak, X. Zarate, J. Tomfohr, O. F. Sankey, A. L. Moore, T. A. Moore, D. Gust, G. Harris, and S. M. Lindsay, *Science* **294**, 571 (2001).
2. T. Ishida, W. Mizutani, U. Akiba, K. Umemura, A. Inoue, N. Choi, M. Fujihira, and H. Tokumoto, *J. Phys. Chem. B* **103**, 1686 (1999).
3. W. Mizutani, K. Tsukagoshi, H. Azehara, and K. Murata, *Jpn. J. Appl. Phys.* **42**, 4535 (2003).
4. Y. Naitoh, K. Tsukagoshi, K. Murata, and W. Mizutani, *e-J. Surf. Sci. Nanotech.* **1**, 41 (2003).
5. Y. Zhang, R. H. Austin, J. Kraeft, E. Cox, and N. P. Ong, *Phys. Rev. Lett.* **89**, 198102 (2002).
6. C. Gómez-Navarro, F. Moreno-Herrero, P. J. de Pablo, J. Colchero, J. Gómez-Herrero, and A. M. Baró, *Proc. Natl. Acad. Sci. U.S.A.* **99**, 8484 (2002).
7. Y. X. Zhou, A. T. Johnson, Jr., J. Hone, and W. F. Smith, *Nano Letters* **3**, 1371 (2003).
8. Y. Yoshioka, Y. Kitagawa, Y. Takano, K. Yamaguchi, T. Nakamura, and I. Saito, *J. Am. Chem. Soc.* **121**, 8712 (1999).
9. A. A. Voityuk, J. Jortner, M. Bixon, and N. Rösch, *Chem. Phys. Lett.* **324**, 430 (2000).
10. F. D. Lewis, T. Wu, X. Liu, R. L. Letsinger, S. R. Greenfield, S. E. Miller, and M. R. Wasielewski, *J. Am. Chem. Soc.* **122**, 2889 (2000).
11. M. Hjort and S. Stafström, *Pys. Rev. Lett.* **87**, 228101 (2001).
12. D. Porath, A. Bezryadin, S. de Vries, and C. Dekker, *Nature* **403**, 635 (2000).
13. M. Amman, R. Wilkins, E. Ben-Jacob, P. D. Maker, and R. C. Jaklevic, *Phys. Rev. B* **43**, 1146 (1991).
14. A. E. Hanna and M. Tinkham, *Phys. Rev. B* **44**, 5919 (1991).

MEDICAL NANOTECHNOLOGY AND DEVELOPING NATIONS

D. C. MACLURCAN[*]

Institute for Nanoscale Technology, University of Technology Sydney (UTS), PO Box 123, Broadway, NSW 2007, Australia

M. J. FORD

Institute for Nanoscale Technology, UTS, PO Box 123, Broadway, NSW 2007, Australia

M. B. CORTIE

Institute for Nanoscale Technology, UTS, PO Box 123, Broadway, NSW 2007, Australia

D. GHOSH

Faculty of Humanities and Social Sciences, UTS, PO Box 123, Broadway, NSW 2007, Australia

The potential for the rapidly emerging field of nanotechnology to assist or further marginalise people in developing nations has been given little consideration. This paper examines the possible need and application of medical nanotechnology, specifically disease diagnosis and drug delivery, in developing nations. It is proposed that the timing may be appropriate for 'intermediate' medical nanotechnology to enter global health discussions. Suggestions for further research and action result from the issues raised.

1. Introduction: Statement of Problem

Although the demand for medical solutions is greatest in poor nations, "only 10% of spending on health research and development is directed at the health problems of 90% of the world's people" [1]. Furthermore, the failure to prioritise health problems has led to the inefficient use of funding [2], whilst official development assistance to Least Developed Countries is decreasing globally [1]. The unstable platform for disease diagnosis and drug delivery is additionally undermined by "outrageously outdated" [3] tools requiring expertise for which there are already depleted human resources [4].

Although revolutionary discoveries surrounding DNA, genomics and proteomics have spurred a greater understanding of the machinations of disease, many benefits have remained limited to populations in the developed nations. Furthermore, the limited success of biotechnology, in raising yield potential of

---

[*] Donald.C.Maclurcan@uts.edu.au

crops [5], has increased disillusion with North-South 'technical assistance'. However, some efforts to bridge the 'digital divide' in the Asia-Pacific region are meeting success [6] and genomics is now considered important enough to be placed on the global health agenda [7]. So, if nanotechnology is the, 'next industrial revolution', why do appraisals of its *global* potential consistently fail to adequately address its possible effects on developing nations?

2. What Area of Application?

The South African Nanotechnology Initiative sees potential for national development of, "...small scale, flexible and low cost technologies", in nanotechnology sectors loosely clustered as 'industrial' or 'social development' [8]. Both fields will generate advancements in areas such as: nanocomposites; molecular electronics; energy efficiency and bioremediation that have much to offer developing nations. However, such applications often require capital-intensive infrastructure for research and design (R&D), production and use, making them more suitable for consideration once 'nanotechnology' and 'developing nations' are used more frequently in the same breath. Although medical nanotechnology may have similar constraints for R&D, some nanomedical devices require minimal infrastructure for diffusion.

Generally speaking, a technology may be deemed appropriate if it can address individual needs, on individual terms. Subsequently, Dayrit & Enriquez suggest biosensors and pharmaceuticals as areas for nanotechnology research based on, "the assumption that most developing economies have significant interests in... public health..." [9]. A University of Toronto study on biotechnology concluded that, 'molecular technologies for affordable, simple diagnosis of infectious diseases' hold the greatest potential for aiding health in developing nations with, 'technologies for more efficient drug and vaccine delivery systems', ranked third [10]. Thus, the authors have broadened potential nanotechnology applications to include diagnostic and drug delivery 'devices'.

3. Addressing Diagnostic and Drug Delivery Problems in Developing Nations Through 'Intermediate' Applications

Despite initially high production costs, prospective estimates are for nanotechnology to enable cheaper, lighter and faster solutions [11] of unforseen proportions. Cost-effectiveness, satisfying Schumacher's first criteria for an 'intermediate' technology [12], stems from novel properties exhibited at the nanoscale. These may be exploited to harness low energy requirements for both production and maintenance [11], exemplified by nanotubes that offer to overcome the wasted heat dilemma of current chip design [13]. Reduced component and production costs may arise from cross-disciplinary collaboration and progress. Large economies of scale are possible with specialised, automated production lines, the use of 'self-assembly' and potential for molecular manufacturing[1] [14, 15]. Ratner and Ratner propose that disease screening could be inexpensive enough to be, "...comprehensive even in low-income countries... as soon as the next two to three years" [13]. Considering that it can cost up to $US15,000 a year to administer antiretroviral drugs to someone in a developing nation [16] the need for more accessible solutions is evident. This example clouds the issue that affordability ultimately depends upon import costs, patent rights and compulsory licensing arrangements. However, technology that is cost-effective from lab to shelf provides a working platform for access issues to be negotiated with greater concern for the recipient.

Nanodiagnostics offer point of care solutions, with heightened sensitivity resulting from selective molecular binding [17], fulfilling Schumacher's second criteria for improved productivity [12]. Prevention, through advanced screening, and response to early detection, are increasing development needs with the emerging prevalence of treatable chronic disease in developing nations.

According to Barton, the ascendency of antimicrobial resistance [18] could be addressed by nanotechnology methods [19]. Freitas' dream of 'magic bullet' drug delivery [20] is growing more realistic through developments in nano-encapsulation and nano-powders for intra-cell delivery, tendering, "...entirely new schemes for increasing bioavailability" [13]. This would in turn lead to reduced dosage requirements and, "...entirely new drugs with fewer side effects and more beneficial behaviour" [13].

The potential for medical nanotechnology to be 'user-friendly' and easily maintained, (therein fulfilling the final component of Schumacher's paradigm [12]), stems from advantages such as non-invasive diagnostic tests using any bodily fluid [21], enhanced time-release mechanisms or the ability to conduct multiple tests simultaneously [22]. Harper sees, "the idea that nanotechnology is all about high technology, semiconductors and science fiction", as a main barrier to the conceptualisation of nanotechnology as appropriate to simple settings [23]. He continues,

> ...near-term applications in nanotechnology, such as drug delivery through the lungs and the skin, have the potential to free up the large numbers of trained medical personnel who are currently engaged in administering drugs via hypodermic needles [23].

---

[1] Though it is not on this speculative basis that nanotechnology is considered in this paper.

Maxmen's 'post-physician era' [24] may never arrive. Yet, the potential for semi-automated nanotechnology devices, to enfranchise the local medic and family administration in developing nations, (especially in rural areas), may be warranted considering the alarming rate at which health care professionals are being lured from developing nations [4].

4. Not Another Technological Panacea: Accepting the Wider Health and Development Context

Amongst many complex problems, including: existing technological dependency; debt; infrastructure and health spending disparities, it easy to see this 'technological panacea' masking a well experienced, "...oppression by industrialized nations..." [25]. This would be based upon labour and research exploitation, as reports point towards a shortfall of nanotechnologists for forecast demands [26]. Furthermore, the few recent assessments of nanotechnology for developing nations have focussed on how the technology *could* be applied without equally contemplating whether it *should*. Ethical implications, extending to health, environmental and cultural concerns, can not be substituted for an 'ends justifying the means' attitude. Perhaps the timing is not prime, for nanotechnology to repeat the mistakes of past technology transfer.

5. Perhaps the Timing is Right?

The World Health Organisation realises the limited resources devoted to health research in developing nations and therein, the vital need to apportion full attention to the most promising technologies [7]. Moreover, Brown sees marginalisation as the end result of the 'development community' ignoring innovations in medicine, "...denying developing countries opportunities that, if harnessed effectively, could transform the lives of poor people and offer breakthrough development opportunities to poor countries" [27]. When considering that nanotechnology is predicted to affect half of the world's drug production by 2011 [28], it is somewhat alarming that the top seven countries produce around 70% of the global scientific papers on nanotechnology. In light of the speed with which nanotechnology is developing, Ratner and Ratner propose that nanotechnology, "could potentially increase the divide between rich and poor nations..." [13]. Yet this divide will only result from inadequate foresight, failure to learn from past technology transfer and poor consideration for current health and development concerns.

On the other hand, Henderson suggests that nanotechnology will, "...open up qualitatively different development paths for some of the developing economies, enabling some regions to "leap frog" their way to leadership" [29].

Ratner and Ratner also consider that, "…if nanotechnology fabrication turns out to be inexpensive and easier to disseminate than current industrial technology… it could reduce the gap between rich and poor or at least make it easier for people around the world to get their basic needs met" [13]. With advances in information and communications technology, bioinformatics and genomics having some influence on foreign aid policies, the collaborative opportunities nanotechnology facilitates [30], suggest opportune timing for this technology to be introduced into global health discussions.

6.   Where To From Here?

Whilst the United States of America, Europe and Japan are at the forefront of nanotechnology R&D, production and policy, the opportunity exists for Australia to establish a benchmark for equitable distribution of nanotechnology potential. Australia's strength in areas of the bionanotechnology sector and reasonably respected position in Asia-Pacific Economic Council (APEC) policy discussions lend themselves to sharing the wealth of knowledge surrounding nanotechnology. Partnership with Australian industry, in some of the poorer regions of the Asia-Pacific, based on strategic importance and competitive advantage [9], must also align with local development and innovation strategies. Whilst it may be, "…imperative that developing economies embark on a nanotechnology S&T program to develop key capabilities and niche areas" [9], private enterprise should only view this opportunity as 'business' if consideration is made for indigenous governance and long-term ownership of public-private partnerships. Should conditionality compromise access to essential services, then the success of medical nanotechnology diffusion at the local level has ultimately failed. An investigation into the transferability of the world's first undergraduate course in nanotechnology, designed at Flinders' University in 1999, may be a platform for the "sharing of national facilities" that Tegart [26] sees as a major need amongst all economies.

Following greater investigation of this papers' considerations and synthesis with the work of the APEC Technology Foresight Network [26], it would be desirable to progress these issues from a nanotechnology-specific conference to a more broad APEC or regional health forum.

7.   Conclusion

Global health problems are compounded by diagnostic and delivery concerns that could be addressed through 'intermediate' medical nanotechnology. Alternatively, the speed with which nanotechnology is growing creates the possibility of a widening health gap. The technologies' potential must be met with concern on an international level, building on the momentum of other emerging technologies. Whilst further research needs to be conducted, Australia is in a position to 'lead the charge' for the equitable distribution of nanotechnology through work in the Asia-Pacific region.

References

1. United Nations Development Program, Human Development Report 2003, Oxford University Press, New York, (2003).
2. Global Forum for Health Research 2003, Priority setting methodologies (Online), Accessed on: July 10 2003. Available: http://www.globalforumhealth.org/pages/index.asp.
3. C. Nacy in V. Brower 2002, Second class medicine for developing nations? (Online), Accessed on: September 16 2003. Available: http://www.vaccinationnews.com/DailyNews/December2002/SecondCl ass6.htm.
4. T. Pang, M. A. Lansang, A. Haines, British Medical Journal, 324, 499-500 (March 2, 2002).
5. V. W. Rutten edited by Laura Engelson 2000, "Biotech has not made impact yet", The Economist, vol. 357, no. 8197, November 21.
6. F. Chand 2003, Annan calls for plan to bridge digital divide (Online), Accessed on: October 28 2003. Available: http://www.rediff.com/money/2003/jul/21un.htm.
7. World Health Organisation, Genomics and world health: report of the Advisorary Committee on Health Research, Geneva, (2002).
8. South African Nanotechnology Initiative, "National Nanotechnology Strategy: Nanowonders -Endless Possibilities, Volume 1, Draft 1.5" South African Nanotechnology Initiative and the Department of Science and Technology,(2003).
9. F. M. Dayrit, E. P. Enriquez 2001, Nanotechnology Issues for Developing Economies (revised) (essay), Philippines.
10. University of Toronto Joint Centre for Bioethics, "Top 10 Biotechnologies for Improving Health in Developing Countries" (2002).
11. The Institute of Nanotechnology, Nanoforum European nanotechnology gateway "Sustainable Production: The Role of Nanotechnologies", Copenhagen (2002).
12. E. F. Schumacher, Small is Beautiful: a study of economics as if people mattered, Blond & Briggs, London, (1973).
13. M. A. Ratner, D. Ratner, Nanotechnology : a gentle introduction to the next big idea, Prentice Hall, Upper Saddle River, NJ, (2002).
14. Center for Responsible Nanotechnology 2003, CRN Research: Current Results — Benefits of Molecular Nanotechnology (Online), Accessed on: August 21 2003. Available: http://www.crnano.org/benefits.htm.

15. The Foresight Institute 2003, Nanotechnology: the Coming Revolution in Molecular Manufacturing (Online), Accessed on: October 15 2003. Available: http://www.foresight.org/NanoRev/.

16. D. Fishburn, S. Green (eds) 2002, "The World in Figures: Industries", The Economist: The World in 2003, p. 102.

17. AMBRI™ Ltd 2003, Point of Care (Online), Accessed on: September 6 2003. Available: http://www.ambri.com.au/Content/display.asp?screen=134.

18. World Health Organisation 1999, Infectious diseases are the biggest killer of the young (Online), Accessed on: 15 July 2003. Available: http://www.who.int/infectious-disease-report/pages/ch1text.html#Anchor5.

19. J. Barton 2001, Antibiotic applications of nanotechnology (Online), Accessed on: August 21 2003. Available: http://web.grinnell.edu/courses/sst/f01/SST395-01/PublicPages/Perfect%20Bodies/Nano/antibiotics.html.

20. R. A. Freitas, Jr., Nanomedicine, Volume I: Basic Capabilities, Landes Bioscience, Georgetown, TX, (1999).

21. AMBRI™ Ltd 2003, Biosensor Movie (Online), Accessed on: October 28 2003. Available: http://www.ambri.com.au/Content/display.asp?screen=174.

22. AMBRI™ Ltd 2003, Products (Online), Accessed on: September 6 2003. Available: http://www.ambri.com.au/Content/display.asp?screen=150.

23. T. Harper 2003, Nanotechnology in Kabul? Taking the first steps (Online), Accessed on: October 10 2003. Available: http://www.nanotechweb.org/articles/column/2/8/2/1.

24. J. S. Maxmen, The post-physician era : medicine in the twenty-first century, Wiley, New York, (1976).

25. H. Lovy 2003, Do they know it's nanotime at all? (Online), Accessed on: October 8 2003. Available: http://nanobot.blogspot.com/106036458329586482.

26. G. Tegart, "Nanotechnology The Technology for the 21st Century" APEC Center for Technology Foresight,(2001).

27. M. M. Brown, in Human Development Report: Making new technologies work for human development United Nations Development Programme, Ed. Oxford University Press, New York, (2001).

28. D. A. LaVan, R. Langer, in Societal Implications of Nanoscience and Nanotechnology: NSET Workshop Report, edited workshop report, M. C. Roco, W. S. Bainbridge, Eds. National Science Foundation, Arlington, Virginia., (2001) pp. 79-83.

29. R. Henderson 2002, The Next Technological Revolution: Predicting the Technical Future and its Impacts on Firms, Organisations and

172

        Ourselves (Online), Accessed on: 16 August 2003. Available:
        mitsloan.mit.edu/50th/tech.pdf.

30.   C. A. Haberzettl, Nanotechnology, 13, R9-R13 (4 July, 2002).